T0213997

SpringerBriefs in Fire

Series Editor

James A. Milke, University of Maryland, College Park, MD, USA

SpringerBriefs in Fire presents concise summaries of cutting-edge research and practical applications across a wide spectrum of fire-related research. Featuring compact volumes of 50-to-125 pages, the series covers a range of content from professional to academic.

Typical topics might include:

- A timely report of state-of-the art analytical techniques
- A bridge between new research results, as published in journal articles, and a contextual literature review
- A snapshot of a hot or emerging topic
- An in-depth case study or clinical example
- A presentation of core concepts that students must understand in order to make independent contributions

Briefs allow authors to present their ideas and readers to absorb them with minimal time investment. Briefs will be published as part of Springer's eBook collection, with millions of users worldwide. In addition, Briefs will be available for individual print and electronic purchase. Briefs are characterized by fast, global electronic dissemination, standard publishing contracts, easy-to-use manuscript preparation and formatting guidelines, and expedited production schedules. We aim for publication 8-12 weeks after acceptance. Both solicited and unsolicited manuscripts are considered for publication in this series. Please email your proposal to: Hector Nazario, Associate Editor

More information about this series at http://www.springer.com/series/10476

Jozef Štefko · Anton Osvald ·
Linda Makovická Osvaldová · Pavol Sedlák ·
Jaroslava Štefková

Model Fire in a Two-Storey Timber Building

Jozef Štefko
Department of Wood Constructions
Technical University of Zvolen
Zvolen, Slovakia

Linda Makovická Osvaldová
Department of Fire Engineering
University of Žilina
Žilina, Slovakia

Jaroslava Štefková
The Institute of Foreign Languages
Technical University of Zvolen
Zvolen, Slovakia

Anton Osvald
Zvolen, Slovakia

Pavol Sedlák
Department of Wood Constructions
Technical University of Zvolen
Zvolen, Slovakia

ISSN 2193-6595   ISSN 2193-6609 (electronic)
SpringerBriefs in Fire
ISBN 978-3-030-82204-0   ISBN 978-3-030-82205-7 (eBook)
https://doi.org/10.1007/978-3-030-82205-7

This Springer imprint is published by the registered company Springer Nature Switzerland AG
The registered company address is: Gewerbestrasse 11, 6330 Cham, Switzerland

We would like to express our special thanks of gratitude

to company FIRES s. r. o. Slovakia.

Foreword

Wood has been a favoured construction material from the beginning of civilization because of its abundance, high stiffness and strength-to-weight ratios and the relative simplicity with which it can be adapted to use. It is therefore not surprising that wood has become an important element in sustainable and economic development and has attracted worldwide attention in recent years.

However, the combustibility of wood is one of the main reasons that too many building regulations and standards strongly restrict the use of wood as a building material. Fire safety is an important contribution to feeling safe and an important criterion for the choice of materials for buildings. The main precondition for increased use of wood in buildings is adequate fire safety.

Worldwide, several research projects on the fire behaviour of wood in buildings have been conducted over the past decades, providing basic data and information on the fire safe use of wood. The current improved knowledge in the area of fire design of wooden structures, combined with technical measures, particularly sprinkler and smoke detection systems, and well-equipped fire services, allow the safe use of wood in a wide field of applications. As a result, many countries have started to revise their fire regulations, thus permitting greater use of wood in buildings. Fire test and classification methods have been harmonised in Europe, but regulatory requirements applicable to building types and end users remain on national basis. Although these European standards exist on the technical level, fire safety is governed by national legislation and is thus on the political level. National fire regulations will therefore remain, but the new European harmonisation of standards will hopefully provide means of achieving common views.

Research on fire performance of wood products has been performed in Eastern Europe for a long time, focusing mainly on material properties, but the contacts with Western Europe have unfortunately been limited, mainly due to language differences. The situation started to change when Professor Anton Osvald from the Technical University of Zvolen in former Czechoslovakia organized the first international conferences on Wood and Fire Safety in 1988. It was, and still is, the only international fire conference dedicated specifically to wood and wood products. It has always been held at Strbske Pleso in the beautiful High Tatra mountains in present Slovakia. I remember long train travels from Bratislava and once even a fantastic helicopter ride with some colleagues.

The conference attracted initially mainly attendants from Eastern Europe. The number of persons attending from other parts of the world has increased gradually and reached a peak in the web-based conference in 2020, organized by Linda Makovicka Osvaldova.

During the conference 2012, a fire test of a whole wooden building was performed in order to demonstrate its fire safety and with the conference delegates invited, including myself. It was also the first main focus on the use of wood in buildings during the conference series. The event is now documented in this book—*Model Fire in a Two-Storey Timber Building*.

It is my sincere hope that the fire research in Eastern Europe will focus more on building applications in order to support the use of larger wood buildings as a sustainable, ecological, esthetic and economic development. I also hope that the mutual research exchange will increase and that more research results from Eastern Europe will be published in international research media.

May 2021

Birgit Östman
Linnaeus University
Växjö, Sweden

Preface

Wood as a building material has an undoubtedly unique position due to its mechanical, thermal and technical, aesthetical, practical and technological properties and the impact on the environment. It is starting to be considered as the raw material of the twenty-first century. In many countries of the European region, it is a strategic raw material while renewable, which brings considerable revenue to national economies, especially there where it is processed in a complex way into products with high added value, such as in timber structures.

The year 2012 brought a significant breakthrough. The technologies of cross-laminated timber, heavy timber constructions and hybrid constructions have come into use; the Eurocodes have started to be fully implemented, and consequently, a multi-storey building made of timber has been approved even in conservative countries. A significant approach shift has taken place when wood—although combustible material—has started to be respected as an alternative material even in multi-storey buildings.

Among the complex fire protection measures, this particular property does not have to play a significant role. On the contrary, the fire resistant properties of heavy wooden profiles, including the strength under a high thermal load, can demonstrate its advantage. Leading research institutes, universities and commercial companies work on research and development focusing on the elimination of risks resulting from the use of wood in the building industry. The solution of this challenge presents multi-storey buildings made of timber which are appearing on numerous occasions all over the world. One of them which is currently taking the attention of the Europeans is the construction of 24 storey timber structure of "Wooden Tower" in Vienna. With a height of 84 m, it will be the tallest wooden skyscraper in the world and will leave other similar buildings behind as for example 10-storey building in Melbourne or 14-storey building in Bergen.

This book, *Model Fire in a Two-Storey Timber Building*, is dedicated to architects, wooden building and construction designers, authorities in fire certification and reviewing, students of master's degree and doctoral students in fire safety, as well as researchers dealing with fire performance of buildings.

Zvolen, Slovakia   Jozef Štefko
Zvolen, Slovakia   Anton Osvald
Žilina, Slovakia   Linda Makovická Osvaldová
Zvolen, Slovakia   Pavol Sedlák
Zvolen, Slovakia   Jaroslava Štefková
May 2021

About This Book

Having accepted the state-of-art knowledge, valid legislation, building materials tested by novelty methods, the model fire of a multi-storey building was prepared. The results of the experiment described in this book, *Model Fire in a Two-Storey Timber Building,* show that the wood-based buildings are equally safe in case of a fire as the buildings built of other materials.

In the introductory chapters, the book presents a complex understanding of fire safety of wooden buildings, discusses the properties of wood so that it is clear how a fire originates and develops in wooden construction, how heat affects the degradation of wood, especially its mechanical properties, and the methods of protection against the fire. Further, fire resistance of wood and wood-based construction assemblies is discussed. It also presents the state-of-the-art knowledge in fire engineering focusing on wooden constructions summarised in Eurocodes. The following part presents the report on model fire of a two-storey building, built in the accredited testing facility FIRES s.r.o. in Batizovce, Slovakia during 7th International Scientific Conference Wood and Fire Safety 2012.

The experiment with complex temperature monitoring in the origin of fire and other hundreds of spots on the surface and inside the constructions in the complete space of the building brought valuable findings about the performance of the fire-loaded wooden building. The model building was constructed of different assemblies based on mineral wool, fibreboard and foam materials, in particular, polystyrene.

The event was accompanied by great interest of the community of professionals and representatives of the main national media and provoked changes in the national legislation.

Contents

Acronyms and Abbreviations

A	Thermal conductivity (W/m.K)
C	Automatic closing
C1-S25	Thermocouples (C1, C2, C3, C4, N45, N46, S49, P4, P48, N19, N20, N27, N42, S21, S22, S23, S24, S25)
c_d	Specific heat (kJ/kg)
E	Integrity (–)
EWS	External Wall Insulation Systems (–)
FSP	Fibre Saturation Point (–)
G	Carbon black burn-out resistance (–)
H	Thickness of wood layer that is thermally stressed (0,04 m/mi)
H_p	Calorific value (MJ/kg, MJ/m^3)
I	Thermal insulation (–)
K	Protection against fires of coatings (–)
$k1$	Coefficient of calorific value reduction due to real fire conditions (–)
λ	Heat conductivity (W/m.K)
M	Mechanical action (–)
MOE	Modulus of Elasticity (MPa)
MOR	Modulus of Rupture (MPa)
Q_c	Overall heat transfer (J)
Q_{cv}	Heat released to the outside space by convection (J)
Q_r	Heat released to the outside space by radiation (J)
Q_s	Heat that causes structure warming-up (J)
$\theta_{S,1}$	Front wall temperature (°C)
$\theta_{S,2}$	Back wall temperature (°C)
$\theta_{S,ex}$	Front wall temperature after exothermic reaction (°C)
θ_3	Ambient temperature (°C)
θ_z	Source temperature (°C)
R	Load-bearing capacity and stability (–)
S	Smoke tightness (–)
T	Temperature under the charred layer (°C)

T_i	Fire ignition temperature of wood (depends on wood species) (°C)
T_p	Charring temperature of wood (°C)
W	Insulation (controlled by radiation) (–)
X, Y, Z	Sensors position (–)

List of Figures

List of Tables

Chapter 1
A Building and a Fire

Abstract Fires have accompanied mankind since the beginning of history. Fires have destroyed unique buildings (the Temple of Artemis in Ephesus, Notre Dame in Paris) as well as whole cities or parts of them. People's efforts to prevent and fight fires are represented in history in various forms such as legal regulations, special fire regulations, and ordinances, and past and ongoing research into the origin and development of fires. Such research monitors the influence of the building materials, construction, and a number of other factors on the origin and development of fires. Individual fires are recorded and various statistics are compiled in order to better understand the origin and development of fire.

Keywords Fire · Building · Fire statistics · Fire origin · Fire development

Structural fire is a phenomenon, which accompanies all forms of civilization. At all places, where civilizations were formed and where people concentrated in small areas, certain forms of fire protection had been created. The fire is not only associated with the entire timeline of a mankind, but also with all people's activities.

From the time point of view, only after many years, people realized that successful firefighting is possible by the concentration in associations, by more effective equipment and by introducing fire protection regulations by governors. For the former Ugrian region, Joseph II the Emperor signed the first fire protection order in 1788, and other reference amendments were created later. There were 25 paragraphs in the order, where obligations were described for house owners, but also for chimney-sweepers, bricklayers and other tradesmen [7]. This order is considered to be the very first standard of firefighting measures in the region. Despite the improved legislation, new materials, advanced technologies, strict administrative arrangements and training, the fire is still around and cannot be omitted nowadays.

The partial answer to the fire cause problem can be solved, if we focus on the time of its ignition during a day.

The relevant statistics for 2020 is therefore shown in Fig. 1.1 (a fire ignition hour during a day) [11]. The data were supplemented (Fig. 1.2) by other relevant fire statistics by Tovey for the entire USA during the years 1970–1979 [12] and fire statistics from the yearly book of Fire Protection of the Slovak Republic in 2000 [8], 2010 [9], 2015 [10], 2020 [11]. If the graph curves are compared, for 5 cases,

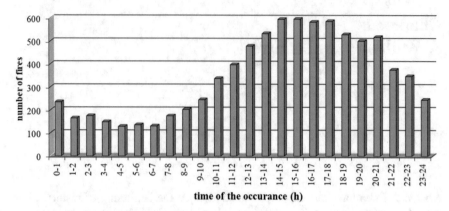

Fig. 1.1 Number of fires per hour during a day, Slovak Republic, 2020

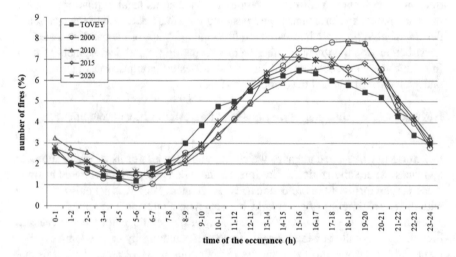

Fig. 1.2 Number of fires per hour during a day for selected years

2 countries and for a single country and different years, it is the visible shape of the curves is very similar.

The number of fires is expressed as percentages in the statistics. The essential information is the shape of the curve and its relation to time. This shape expresses that it is mostly people who are to blame for fires, and not the materials, products or buildings, which are equally combustible at any hour of the day. If their properties posed the main problem, the curves would have a linear character.

This conclusion points to an idea of using combustible structural materials in the right way without increasing the fire risk, meaning that we need to understand the behavior of combustible materials during a fire in detail, and use high quality and certified materials and good workmanship during installation.

Fig. 1.3 Overall heat balance in an enclosed space

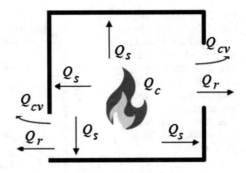

Fires in buildings cost millions of dollars every year and turn them into ruins in minutes. Moreover, if such accidents claim lives, the consequences are much more significant; people cannot be replaced, and those wounded spend many days in pain and often sustain lifelong injuries.

In the case of a fire in an enclosed space—in a building (Fig. 1.3)—the fire energy (heat Q_c), the heat produced by the fire divides into the structure heating Q_s, and the heat released and spread by the fire itself by convection Q_{cv} or radiation Q_r.

If we assume the fire is isolated from other spaces by a sufficiently large space with no flammable materials (therefore without fire risk), and the energy of the radiant component of thermal energy cannot heat the materials behind the space without fire load to the material's ignition temperature, then the natural consequence will be that the conditions for burning propagation will be unfavorable, and the fire will not continue. A properly designed composition of ceilings and walls, as well as multi-story wooden buildings, can ensure these conditions, which prevent the spread of fire outside the area of its origin [1, 2, 4].

The equation of the total heat balance can be expressed as follows according to Formula 1.1:

$$Q_c = Q_s + Q_{cv} + Q_r \text{(J)} \qquad (1.1)$$

where:
Q_c—overall heat flux of a fire, or total energy (J);
Q_s—heat that causes heating of a structure (J);
Q_{cv}—heat released by convection (J);
Q_r—heat released by radiation (J).

Wood, plastics, and other advanced synthetic materials that are used in building structures or the interior furnishings can, during incomplete combustion, create large amounts of dangerous smoke. Long-term, targeted international research has shown that in addition to suffocating and toxic effects, the content of unburned decomposition products in smoke have a much greater impact on the course of a fire than

Fig. 1.4 Fire temperature curve [27] 1—linear form of fire propagation (current modern presentation of the time–temperature curve), 2—nonlinear form of fire propagation ("historical" presentation of time–temperature curve), 3—possibility of "back draft" effect

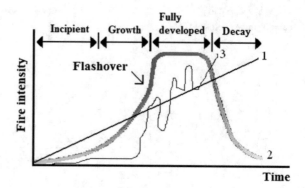

previously thought. This was also reflected in the modern interpretation of the time–temperature course of the fire. Each fire is characterized by a time–temperature curve (Fig. 1.4), which can be divided into four separate stages.

The first stage of a fire is its initiation. The initial fire is the beginning of the burning process and the fire's formation, in which the main event is the ignition of individual materials and their reaction to the sources of ignition. In the first stage, after the combustion process starts, the speed of flame propagation on the surface, the calorific value of the material, and the material's contribution to further combustion are important. This stage depends on the method of energy release, the type of fuel, and the characteristics of the structure delineating the space in which the fire occurs. The materials' retardation, or their treatment against slow and fast ignition, is of great importance at this stage. Here is the decisive moment which determines whether a fire will occur at all [1, 3, 5, 6].

In addition to the material, the course of the fire is also influenced by the space where it is burning, especially the ventilation and airflow in the fire area. This supplies one of the materials which is of utmost importance for the whole combustion process—oxygen. If the material components (material, oxygen) and energy component (heat) are balanced in favor of combustion, even if imperfectly balanced, the second stage of fire occurs—a developed fire.

A special turning point of the second stage of the fire is "flashover," or spatial ignition. The emergence of the flashover effect in practice means not only the end of the second stage of fire development and the very fast transition or "jump" to the third stage, a fully developed fire. At the same time, flashover is the beginning of the critical thermal stress of the structural elements, as a result of which the statics can be disturbed and the building can collapse. Flashover is a situation where all the combustible materials in the whole fire space in an enclosed room ignite rapidly at once. The main cause of flashover (spatial ignition) is the simultaneous heating, by combustion products, of all combustible materials in the room to a temperature at which flammable gases and vapors are released. If the accumulated gases and vapors reach the lower flammability limit and the room temperature is high enough (500–650 °C), they ignite immediately. Modern test methods which evaluate materials' reaction to fire categorizes materials in terms of fire protection, so that if they meet fire reaction classes A1, A2, or B, then flashover/spatial ignition, and consequently the third stage of fire development, cannot occur, and the fourth phase (fire decay) follows immediately after ignition [13].

The third stage of a real fire is a fully developed fire (see Fig. 1.4). In this stage of the fire, the fire spreads outside the premises where it originated. Fire resistance tests of the structures are adapted to the conditions of this stage of fire.

The fourth stage of the fire is the decay or extinguishing of the fire, either by means of intentional fire-fighting or by self-extinguishing if all the fuel has burned away.

Fires are also changing and evolving. Two minutes after ignition, the average fire has a higher temperature than it had 20 years ago. The reason is the composition of the materials used as furnishings. In 1950, the temperature of an apartment fire peaked after 15 min at 760 °C. Today, it can reach 1100 °C in a mere three minutes. The fire grows from a seemingly harmless flame—and the energy released during the fire can destroy the apartment in less than ten minutes.

References

1. Balog K (1999) Samovznietenie (Self-ignition). SPBI, Ostrava. ISBN 80–86111–43–1
2. Buchanan A, Abu A (2017) Structural design for fire safety, Second edition. John Wiley & Sons Ltd, West Sussex. ISBN 978-0-470-97289-2
3. Giertlová Z (2009) Brandsicherheit von mehrgeschossigen Wohnge-bäuden in Holzbauweise (Fire safety of multistory timber residential building), habilitation presentation 10.12. 2009. Faculty of Safety Engineering,VŠB–Technical University of Ostrava
4. Karlsson B, Quintiere JG (2000) Enclosure fire dynamics. CRC Press, London. ISBN 0-8493-1300-7
5. Mikolai I, Tkáč J (2007) Príjazdové komunikácie a nástupné plochy dustup-nosť budov. (Arrival communications and fire lines–accessibility of buildings) In: 31. International scientific conference of civil engineering departments, 2007: Proceedings, Čeladná, ČR, 19.-1.9.2007. VŠB – Technical University of Ostrava, Ostrava. pp 147–152. ISBN 978–80–248–1405–6
6. Mikolai I (2009) Dvere na únikovej ceste–návrh a realizácia v praxi. (Doors for emergency exit – design and execution in the practice) In: FIRECO 2009: VIII. International conference, Trenčín, SR,13.-14.5.2009. Požiarnotechnický a expertízny ústav MV SR Bratislava, Bratislav. ISBN 978–80–89051–10–6.
7. Osvald A, Balog K. (2017) Horenie dreva. Zvolen: Vydavateľstvo Technickej univerzity vo Zvolene. pp 106. ISBN 978–80–228–2953–3
8. Ročenka požiarnej ochrany (2000) The yearbook of fire protection 2000. Úrad požiarnej ochrany Ministerstva vnútra SR, Bratislava
9. Ročenka hasičského a záchranného zboru (2010) The yearbook of fire and rescue corps 2010. Ministerstva vnútra SR, Prezídium Hasičského a záchranného zboru, Bratislava
10. Ročenka hasičského a záchranného zboru (2015) The yearbook of fire and rescue corps 2015. Ministerstva vnútra SR, Prezídium Hasičského a záchranného zboru, Bratislava
11. Štatistická ročenka 2020 HaZZ (2020) The yearbook of statistics of fire and rescue corps 2020. Ministerstva vnútra SR, Prezídium Hasičského a záchranného zboru, Bratislava
12. Tovey H (1997) The collection and uses of fire data. Wood Fiber 9(1):44–59
13. Troizsch J (1966) Fire regulations and testing of building materials in Europe: status and perspectives. In: Conference proceedings: wood & fire safety. Technická univerzita vo Zvolene, Zvolen, pp 11–20. ISBN 80–228–0493–2

Chapter 2
Wood—A Combustible Building Material

Abstract This chapter describes wood as a combustible material, discussing the influence of thermal degradation on individual wood structures (microscopic and macroscopic) as well as on changes in its chemical composition and changes in physical and mechanical properties.

Keywords Fire · Wood · Thermal degradation of wood · Changes in wood

2.1 General Characteristic of Wood

Wood has technical properties that predispose it to general use. It is used to make objects of all kinds, buildings, furniture, musical instruments, and objects of an artistic nature. Unique historical monuments of wood have been preserved until today, such as statues, altars, carved furniture, as well as entire wooden buildings, churches, bell towers, gates, parts of castles and chateaux and entire wooden villages [1].

The gradual deepening of knowledge about the internal structure, chemical composition, and physical and mechanical properties of wood goes hand in hand with the intensive development of technologies for wood processing and its versatile use. It is relatively simple and inexpensive to process. It can be glued, modified, and joined with or without fasteners or other joining materials [2].

In the past, the primary and only function of wood was to serve as fuel. Only later did wood begin to serve other purposes, such as a building material and material for various types of products, such as objects of daily use or of artistic value.

Wood in any form (raw material, semi-finished product, or end product) is a combustible material. However, its flammability can be regulated in various ways. Adjustment of the fire parameters of wood is necessary mainly for wooden buildings.

There are several ways to regulate the origin and development of fires in wooden buildings. The basic methods are given in Chap. 1 and are valid for all types of constructions, not only wooden ones. The application of the basic building material (wood or wood-based materials), its properties, method of use, and design can also significantly affect the behavior of the final product—the wooden building—in the case of a fire.

In this chapter, we will outline various new findings on the behavior of wood and wood-based materials at higher temperatures. This is only an overview, not an overall analysis, which is currently quite detailed and is also applied in the latest regulations. We should use this new knowledge in practice to be able to build wooden buildings that are safe in all respects [3].

2.2   Wood Degradation by Heat Impact (Fire)

2.2.1   Thermal Degradation of Main Wood Components

Wood is a biopolymer material with a high energy content, which accumulated in its chemical structure by means of photosynthesis (the formation of glucose from carbon dioxide and water) and consecutive endothermic chemical reactions (transformation of glucose into polysaccharides and lignin). This energy can be reversibly released by thermal activation.

This means that if an effective activation source (flame, radiant body, etc.) starts to act on the wood, the opposite process occurs, i.e., that the basic building elements of wood (hemicellulose, cellulose, and lignin) with a high energy content are broken down to form flammable gases. At sufficiently high temperatures, flammable gases react with oxygen. Carbon oxides and water are formed in various thermal oxidation reactions of an exothermic nature, a considerable amount of energy is released, which participates in further heating and pyrolysis of wood. Combustible gases can ignite at some point (usually at temperatures around 250 °C) and at this point, the wood can enter the spontaneous combustion phase, i.e., without the need for an external heat source [4].

Wood burning can be characterized as a chemical reaction where gas products produced during primary thermolytic decomposition of polysaccharides and lignin (depending on wood type and auxiliary substances) react with oxygen in an exothermic reaction while generating thermal and light energy.

Thermal decomposition of wood is a set of chemical reactions initiated by activating thermal energy. This causes excitation of electrons in the covalent bonds in the polysaccharides and lignin to a higher energy level. At temperatures below 66 °C, these reactions do not take place in principle; at temperatures from 66 to 110 °C some may take place (depending on the duration of the heating), but they are of negligible impact on the structure and properties of the wood. More obvious thermal degradation of the building components of wood occurs only at temperatures above 150 °C, when hemicelluloses, cellulose, and lignin decompose. Based on thermal and physicochemical analysis, it can be stated that [5]:

– Hemicelluloses are the most thermally vulnerable wood component. Their thermal decomposition occurs in a wide temperature range, from 170 to 240 °C (with a more significant exothermic effect above 200 °C),

- Cellulose is more thermally stable compared to hemicelluloses; up to 250 °C the decomposition of cellulose is slight; significant depolymerization occurs at temperatures above 300 °C, when bonds in the base of the polymer chain start to break up and the terminal link transforms to levoglucosan, which is subsequently converted into flammable gases,
- Lignin is thermally the most durable wood component, as its 3D benzenoid structure is highly resistant to heat; heating up lignin (100–180 °C) causes its plasticization (endothermic phase); significant exothermic decomposition occurs from 300 to 400 °C in relation to bond breaking in its aromatic ring.

Increased temperatures above 100 °C affect changes in the physical, structural and chemical properties of wood. Besides temperature, there are also other factors affecting the changes—time, atmospheric pressure, and water content—so under certain conditions, changes in wood can be observed even at temperatures up to 100 °C [6–9].

2.2.2 Wood Microscopic Structure Changes by High-Temperature Stress

The anatomic structure, or individual cell elements, affects the burning process, mainly the first phase—the ignition. This is determined by chemical components and by the geometrical shape of cell elements, by their dimensions and number.

These statements are documented by studies using scanning electron microscopes, shown in Fig. 2.1 [1].

The figures show the boundary of the annual ring of spruce wood before thermal degradation (2.1 a) and after thermal degradation (2.1 b). After thermal degradation,

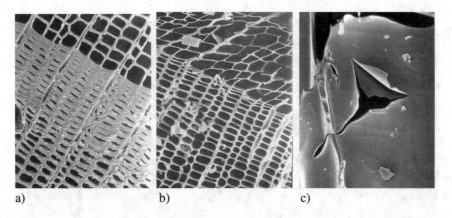

a)   b)   c)

Fig. 2.1 Scanning electron microscope [1]

the main thinning of the cell wall can be seen; it is caused mainly by degradation of hemicelluloses and cellulose in the secondary layer of the cell wall.

The lignified middle lamella and the primary wall retain their shape even after thermal degradation, even just before incineration. This can be seen in Fig. 2.1c, (detail) which shows the division of the two primary layers of the cell wall in the middle lamella.

2.2.3   Macroscopic Structure Changes of Wood Under High Temperatures

Changes in macroscopic features during thermal degradation are very difficult to evaluate. If macroscopic changes are observed in the first stages of thermal degradation (very high temperatures and a very short time frame—up to 1 min, or in thermal degradation at lower temperatures), cracks are observed. These are basically dry cracks, which also occur during poor and fast drying of wood. As the degradation progresses, the cracks enlarge and the wood begins to cube-crack. The surface exposed to the flame increases and the combustion intensifies.

Thermal degradation also changes the color of the wood (Fig. 2.2) [1]. The light color of the wood gradually darkens until it changes to a black-charred layer of wood. The intensity of darkening depends on heating rate and temperature. The color change is caused by the degradation of the polysaccharide content (hemicelluloses and cellulose). In all cases, the final phase is the formation of a carbonized layer. The carbonized layer is black—a good absorber of thermal radiation. However, due to its chemical composition and porous structure, it is also a poor conductor of heat. Therefore, it is customary to talk about the auto-retardation character of wood in connection with the charred layer, which is also taken into account in practice.

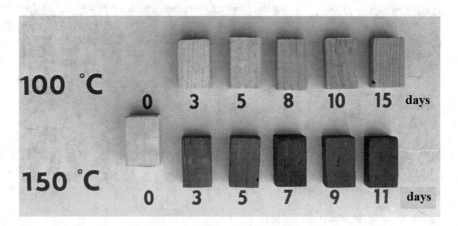

Fig. 2.2 Wood color change due to thermal degradation [1]

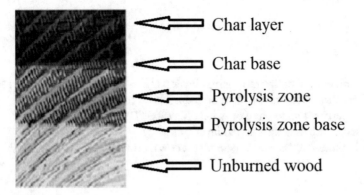

Char layer

Char base

Pyrolysis zone

Pyrolysis zone base

Unburned wood

Fig. 2.3 Charred layers of wood

When observing macroscopic changes during the advancing stages of thermal degradation at high temperatures over a longer period of time, only the formation of a carbonized layer can be observed. First to appear is a homogeneous black thin layer that gradually thickens and cracks cubically. After cracking, it peels off, and changes the cross-section of a thermally degraded wooden element.

The carbonized layer covers the entire surface of the wood. If the element is cut, the individual layers can be observed, as shown in Fig. 2.3. The carbonized layer is not homogeneous but consists of thinner films—each having different properties created at a different temperature.

Interesting research was performed by Osvald and Balog [1], who monitored the change in temperature in the cross-section of the beam directly during the fire resistance test. As the wood burned, the characteristics of the cross section changed until the beam finally lost its load-bearing capacity. After removal of the charred layer, the thermally degraded layer of wood can be seen—it can be detected visually by the color change. However, this layer is only 11–15 mm thick and there is completely intact wood behind this layer (Fig. 2.4).

The integrity of the wood is indicated not only by its color but also by the tests performed on such wood. These were tests of chemical analysis, percentage determination of the basic components of wood (hemicelluloses, celluloses, lignin) as well as determination of their quality. Mechanical and physical tests were also performed and confirmed that the wood retained its parameters without failure. This gives designers a chance to increase the fire resistance of solid or glue laminated timber beams by means of carbonized layer allowance, and this principle is beginning to be applied in Eurocode 5 [10]. The temperature under the charred layer can be calculated according to the following Eq. (2.1):

Fig. 2.4 Wood color changes by the thermal degradation [1]

$$T = T_i + (T_p - T_i)\left(1 - \frac{x}{h}\right)^2 (°C) \tag{2.1}$$

where:

   T—temperature below the charred layer (°C);

   T_i—ignition temperature of wood (depends on tree species) (°C);

   T_p—charring temperature of wood (°C) (300 °C);

   x—distance below the charred layer (mm);

   h—thickness of the wood layer exposed to thermal stress (0.04 m/min).

2.2.4 Physical Properties of Thermally Degraded Wood

In addition to chemical composition, the physical properties of wood and wood-based materials significantly affect the burning process. Each of the physical properties has an effect on combustion, although not all of them have the same effect [1].

The structure of wood and the structure of wood-based materials, in addition to significantly affecting other physical properties of the material, directly affect their combustion. This is due to the size of the openings of the micro- and macro-capillaries, which affect the transport of oxygen into the mass as well as the release of volatile wood products when the conductive elements are open.

The volumetric mass density of wood and wood-based materials is an important property that significantly affects all physical and mechanical properties of wood and also significantly affects the burning process. The density indicates the weight of a specified volume of wood and is most often expressed in kg/m^3. This value is affected by humidity, so it is typically given at a specified moisture level. It is logical that a denser material consumes more energy for ignition and combustion. However, some statements, which differentiate the flammability of individual trees according to their volumetric mass density, are not correct. The chemical composition is a more important criterion than density; for example, woods with a higher hemicellulose content are easier to ignite even at a higher density. With the same chemical composition and the same percentage of the main components of the wood, the effect of density on ignition and combustion is manifested. This effect of density on flammability becomes even more obvious in large wood-based materials.

The surface of the material (its quality) is another physical characteristic that has a significant effect on combustion. Wood, a capillary-porous material, has a level of roughness which, in addition to the method of processing, depends on the anatomical structure of the wood. The surface quality can also be affected by anatomical defects, such as defects occurring during woodworking, mechanical damage, dirt, and other factors changing the surface quality.

The quality of the surface mainly affects the heat transfer and material transfer coefficients. A high-quality smooth surface reflects the energy of a radiant and flame

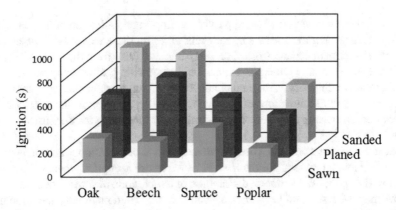

Fig. 2.5 Time-to-ignition of different tree species depending on the surface quality [1]

source and is, therefore, more difficult to ignite than a rough surface under the same thermal load [11].

These conclusions were confirmed by an experiment in which the influence of surface quality on time-to-ignition at a constant heat source was studied. Four wood species (spruce, beech, oak, and poplar) were subjected to the experiment. Three surface types were monitored: sawed, planed, and sanded. As can be seen in Fig. 2.5, the sanded surface was most resistant to the heat source, and achieved the longest times required for the ignition. This factor is so significant that in many countries, sanding beams to a certain smoothness is a standard, not only in terms of aesthetics but also in terms of fire resistance.

Moisture—the water content in wood is a special chapter in the woodworking industry. Moisture affects the volumetric mass density of wood and almost all its other physical and mechanical properties. Monitoring the humidity level forms the basis of technological disciplines in wood processing, including the preparation of semi-finished products and the production of wood products. Wood moisture also affects the burning process.

As the water content of the wood increases, so does the resistance of the wood to ignition. This is explained by the fact that part of the energy is used to evaporate the free water and to break the bonds and evaporate the bound and chemically bound water. Flammable gases diluted with water vapor have a lower concentration and thus lower flammability.

The water contained in the wood acts as a flame retardant, but we cannot attribute the auto-retarding effect to moisture as in the case of the charred layer. Although moisture makes ignition difficult, it has the opposite effect once burning is initiated. The formation of steam in the structure of wood causes the formation of microcracks and cracks; the compactness of the wood decreases, the area exposed to fire increases, and the compactness of the charred layer is disturbed. All of these factors cause more rapid degradation of wet wood, even though initial ignition was problematic.

The thermal properties of wood provide an important indication of the materials, although, regarding their values in the event of a fire, these values are problematic. Out of the thermal properties of wood, it is necessary (also from the point of view of fire protection) to know: thermal expansion, the specific heat capacity of the wood c_d, thermal conductivity λ, thermal conductivity A, and transfer characteristics.

Increasing the temperature of the body causes an increase in the energy of its molecules, increasing their velocity, and ultimately increasing the dimensions of the body.

Heat transfer in wood and wood-based materials is of great importance in terms of fire protection. Knowledge of the laws of heat transfer is applied whenever we need to know the spatial and temporal distribution of temperature in wood. The assessment of the thermal insulation properties of wood, wood-based materials, and structures is also important.

In general, all three basic types of heat transfer can occur in wood. Heat conduction is the form of energy transfer through mass, the elements of which remain at rest. Heat flow (convection) is the transfer of energy through the mass, whose elements are in motion. Heat radiation is the emission or reception of energy in the form of radiation, and no mass is required for the transfer of energy between two bodies. The share of convection and radiation in the total heat transfer in wood is usually small. To simplify the transfer model, the entire energy transfer is described as heat transfer and any deviations are included in the corrections of the transfer characteristics for the given conditions [12].

In the event of a fire, the problem of the thermodynamic properties of wood and heat transfer in wood is even more complicated. We do not know the exact values of the source. We previously mentioned the formation, properties, and function of the charred layer of wood. Its thickness changes and so does the conductivity, which has a positive effect on the evaluated property of the wood.

There is also a negative effect in the initial phase of the burning process and the formation of an exothermic reaction, as shown in Fig. 2.6. This makes other

Fig. 2.6 Heat transfer through wood θ_z—source temperature, $\theta_{S,1}$—front wall temperature, $\theta_{S,ex}$—front wall temperature after the exothermic reaction, $\theta_{S,2}$—rear wall temperature, θ_3—ambient temperature [1]

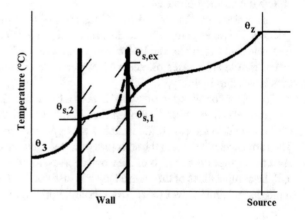

related calculations more difficult to work out according to theoretical relations [1, 2]. Therefore, concepts such as fire calorific value have been introduced into practice.

The heat of combustion is the heat of reaction, i.e. the amount of heat released by the complete combustion of a mass or volume unit of flammable substance in oxygen. The condition is that the water contained in the combustible substance is released in the liquid state. From the point of view of fire protection, the heat of combustion has a decisive influence on the spread of fire, its course, and on the course of temperature increase over time during the fire. The combustion heat of evaluated materials has also become the main evaluation criterion for the classification of material into class A1 and A2 according to STN EN 13,501-1.

*Calorific va*lue is the amount of heat released under the same conditions as the combustion heat. The difference between the heat of combustion and the calorific value is that in the determination of the heat of combustion, the water vapor present in the combustion products condenses, while the water vapor does not when combustion heat is taken into account. It follows that the calorific value is the heat of combustion, reduced by the condensation heat of the water contained in the combustion products.

The calorific value corresponds more to the real combustion conditions, and as such, is the basis for determining the fire load in buildings and for calculating intensity of a fire. The fire load represents the amount of heat that, in the event of a fire, is released on each square meter and subsequently acts on the surrounding area. The reference fire load value is a single kilogram of wood with a calorific value of 16,347 MJ/kg.

In a fire, under normal conditions, the combustion of flammable substances does not usually release such an amount of heat. The energy, released in simulated conditions from a unit of tested material, is expressed by the fire calorific value Hp (MJ/kg, or MJ/m^3 respectively), while created water remains in exhaust gases as water vapour.

The reduction in calorific value under fire conditions is expressed by the coefficient of the fire calorific value $k1$ (-), by means of which the standard value of calorific value is converted into the fire calorific value. For solids that are not sufficiently characterized, approximate values may be used:

$k1 = 0.75$ for permanent fire load,

$k1 = 0.8$ for accidental fire load,

$k1 = 1.0$ for flammable liquids or gases.

The geometric shape of the material significantly affects the possibility of ignition, speed, and intensity of the combustion process. Dimensions, especially thickness, length, diameter, number of edges, angles of rounding, and other parameters of geometric shape determine the resistance of the wooden element to ignition. An important variable is the volume-to-surface area ratio. With the decreasing value of this ratio, as in fragments, chips, dust, wooden objects are very easily ignited. The schematic distribution of the parameters of the geometric shape is shown in Fig. 2.7.

Electric, magnetic, optic or acoustic properties do not have a direct effect on ignition and combustibility of wood and thus do not have the effect on the fire properties.

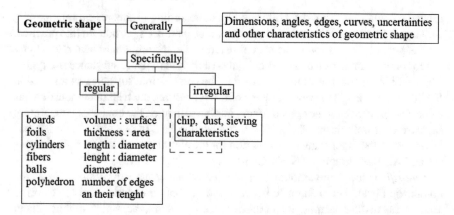

Fig. 2.7 Geometric parameters of wood elements affecting ignition and combustibility [1]

2.2.5   Mechanical Properties of Thermally Degraded Wood

Wood is widely known and used in practice mainly for its unique mechanical properties, especially its elasticity and strength. In combination with a relatively low density, wood has high strength, which predetermines it to be a good structural material.

In this subchapter, we will point out the influence of heat on selected properties (bending, compression, tension, shear), as a comprehensive discussion of every property is beyond the scope of this chapter.

Wood has a special position among building materials with regard to fire-fighting properties. Its load-bearing capacity is not reduced all at once, but gradually, by the burning of the cross-section. For other building materials, these properties have more sudden shifts, e.g. the load-bearing capacity changes suddenly by changing some quantity—specifically by increasing the temperature of the material (e.g. steel).

With a change in temperature, the basic mechanical properties of wood (strength, flexibility, plasticity, and toughness) change.

The increase in temperature may result in temporary changes in the material's properties due to temporary change in internal energy levels, or it can induce permanent changes of the wood's mechanical properties.

If the wood is exposed to higher temperatures for a longer time, its strength is permanently reduced due to degradation of the material, and corresponding weight loss and change of dimensions (cross-section). The magnitude of these effects will depend on the moisture level, method of heating, temperature, exposure time, type of wood, and size of the sample [13].

The mechanical properties of wood up to about 150 °C are approximately linear with temperature. The change in material properties that occurs when the wood is rapidly heated or cooled and then tested immediately is called the immediate effect.

At temperatures below 100 °C, the immediate effect is essentially reversible, and the property returns to the value at the original temperature if the temperature change is rapid [13].

Influence of Temperature on Bending Strength and on Bending Modulus of Elasticity

If we compare the experimental results of different researchers, we should compare only the relative change of the observed characteristics, which is related to the reference temperature of 20 °C. In their experimental work, Glos and Henrici [14] monitored the bend strength and modulus of spruce wood in the temperature range of 20, 100, and 150 °C at an initial wood moisture level of 12%. Figure 2.8 shows the relative bending strength of spruce wood in relation to temperature. The results show that the bending strength decreased by 28% up to a temperature of 100 °C, and the strength decreased by 42% at a temperature of 150 °C. The bending modulus of elasticity decreased linearly with increased temperature. Figure 2.9 shows the relative bending modulus of elasticity of spruce wood as a function of temperature. At 100 °C, the flexural modulus of elasticity decreased by 12% and at 150 °C by 19%.

Fig. 2.8 Relative bending strength of wood in relation to temperature [14]

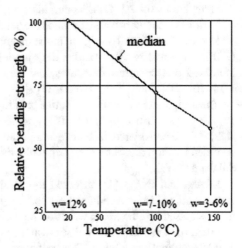

Fig. 2.9 Relative bending modulus of elasticity of wood in relation to temperature [14]

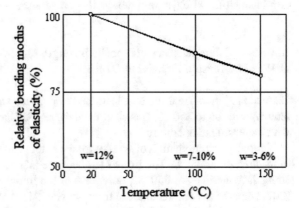

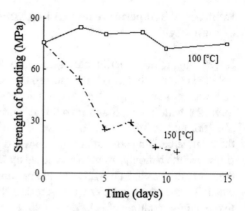

Fig. 2.10 Modulus of Rupture (MOR) at static bending—changes due to different thermal stress and exposure time [1]

These values refer to the values of bend strength and flexural modulus of elasticity at a temperature of 20 °C and a humidity of 12% [3, 15].

Various researchers have examined the bending properties of wood in the temperature range of −140 to + 60 °C, at moisture contents above the Fibre Saturation Point (FSP). They realized that the bending modulus of elasticity and bending strength increases as temperature decreases, and the relationship is linear in the range from −140 to −20 °C and from 0 to 60 °C [3].

The temperature impact on relative bending strength at 0% moisture content [11, 16–18] was described by Gerhards [5], while the reference value of 100% moisture at 20 °C was considered. The lowest impact of temperature on bending strength was recorded by Partl and Strassler [16], while the highest impact was demonstrated by Sulzberger [17].

Osvald and Balog [1] observed the Modulus of Rupture (MOR) of wood at static bending under a radiant heat source (Fig. 2.10). The specimens were thermally stressed and the MOR was examined after cooling. No strength reduction was recorded for long-term thermal stress at 100 °C. The temperature of 150 °C caused a sudden strength reduction. This reduction was so abrupt that the experiment was terminated after 11 days even though the temperature difference was only 50 °C.

Influence of Temperature on Tensile Strength and Modulus of Elasticity of Wood in Tension Parallel to Grain

Östman [19] monitored the tensile strength of spruce wood parallel to the fibers at temperatures up to 250 °C. It is clear from the results that the deformation decreases with increasing temperature.

The tensile strength limit of spruce wood was reduced by 40% at 200 °C compared to its strength at 25 °C. The further temperature increase caused a more significant strength reduction—at 250 °C, only 35% of the initial strength remained. Östman [20] compared her results to those from Nyman [21], Knudsen and Schniewind [22] and Schaffer [4]. Figure 2.11 shows the comparison of tensile strength changes.

Fig. 2.11 Impact of thermal
stress on tensile strength,
parallel to grain [19]

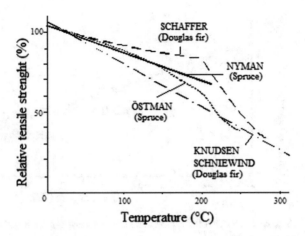

Knudsen and Schniewind [22] did not find a sharp boundary for reduction in these characteristics, at 200 °C for samples with a moisture content of 12%. They state that at a temperature of 250 °C, the tensile strength of spruce wood changed by 40–50%. Nyman [23] obtained a linear relationship with both tensile strength and modulus of elasticity up to 200 °C. Schaffer [24] performed his experimental measurements on zero-moisture samples. His data suggest that at temperatures above 200 °C there is a rapid decrease in tensile strength, which is probably due to rapid thermal degradation.

Influence of Temperature on Mechanical Properties of Wood in Pressure Parallel to Fibers

As the temperature rises, the compressive strength of the wood decreases. The degree of strength reduction at a given temperature depends on the wood's moisture content as well as the exposure time. After 3–4 h of exposure to temperatures around 200 °C, a sharp decrease in strength, mainly caused by the increased intensity of thermal vibration of molecules, is observed [23, 25, 26].

Osvald and Balog [27] came to interesting conclusions after monitoring the compressive strength of spruce wood parallel to the grain. At a temperature of 100 °C, the compressive strength of spruce wood increased, but at a temperature of 150 °C, an irregular reduction in strength was observed. The experiment lasted 15 days.

Osvald [28] found that the compressive strength of spruce wood parallel to the grain at temperatures of 100, 150, 200, and 250 °C at exposure times of 15, 20, 25 and 30 min were as follows. At temperatures up to 200° C, there was no significant difference in compressive strength at different time exposures; the average compressive strength at this temperature was 54.84 MPa. At 250 °C, the time intervals caused differences in compressive strength. The average strength value decreased by 50% compared to the corresponding strength at 20 °C.

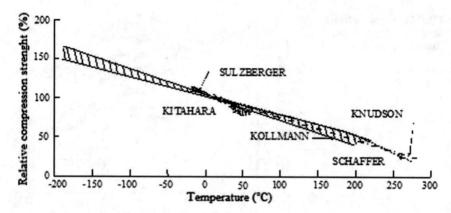

Fig. 2.12 The effect of temperature on compression strength of wood parallel to grain with 0% moisture content [5]

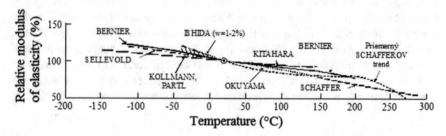

Fig. 2.13 Temperature effect on modulus of elasticity (MOE) of wood parallel to grain with moisture content of 0% [5]

In his work, Gerhards [5] graphically compared the results of research [4, 17, 22, 29] concerning the influence of temperature on the compressive strength of wood parallel to the grain at a 0% moisture level (Fig. 2.12). 100% relative compressive strength is at 20 °C. The results of the authors show relative consistency with each other [24].

Experimental results [3, 11, 30–32] summarized by Gerhards [5] are shown in Fig. 2.13, which illustrates the effect of temperature on the Modulus of Elasticity (MOE) of wood parallel to grain with a moisture content close to 0%. The reference MOE at 20 °C is used.

Influence of Temperature on Shear Strength of Wood

The results of Kitahara and Matsumoto, Sadoh, Young [33–35] and others suggest that the modulus of elasticity in shear (torsion) is particularly sensitive to temperature changes and best reflects the internal changes in wood caused by temperature.

Law a Koran [36] observed tensile shear stresses of wood at various temperatures ranging from 22 to 150 °C. The results show that the rate of reduction of shear stress is significantly higher for hardwoods than for softwoods. According to the results of this study, the maximum shear stress was reduced by 77% in a given temperature range. This reduction is attributed to the thermal plasticization of hemicelluloses and lignin [5]. Stone [37] reports an 80% reduction in shear stress of birch wood in the temperature range of 25–190 °C.

Results confirmed that the selected initial density has an effect on the parameters at all temperature ranges, as the test samples had been carefully chosen by density. Toughness is affected not only by density, but also by exposure time, particularly in denser wood [38].

Influence of Wood Charring Rate on Mechanical Properties in General

The pyrolysis zone is a thin layer of material, only a few millimeters thick, which forms between the charred layer and the wood during the process of burning. When the pyrolysis gases leave the charred surface of the wood, they are subjected to flame burning. Determining the charring rate of wood is crucial because the strength of the charred layer has virtually zero bearing capacity.

Wood generally begins to char around 290 °C. The charring rate of wood depends on the kinetics of wood-burning, including external exposure characteristics. The burning (combustion) of wood is an extremely complex process exacerbated by the development and spread of flame gases.

Decomposition products are dependent on temperature, density, and wood species, while transfer processes depend on wood species, moisture content, permeability, and other morphological factors. At the same time, the thermal properties of wood influenced by these processes are constantly changing, as is the temperature itself [8, 39].

When determining the resistance of wood elements, analysis is based on the reduction of the cross-section of the element as a result of burning. The results of the experimental work of several authors [4, 36, 40] confirm that the burning rate of wood is constant.

Figure 2.14 is a variation of parallel-to-grain strength elastic modulus of wood with temperature according to EN1995-1-2 [41]. The same authors state a variation of parallel-to-grain Young's elastic modulus of wood with temperature according to EN1995-1-2 [41] (Fig. 2.15).

The burning rate is influenced by various characteristics of wood such as density, moisture, permeability, structure, and chemical composition. Bulk density is considered to be a significant factor influencing the burning rate of wood [4]. In general, we can assume that wood that has a higher bulk density usually has a lower burn rate. Increased moisture content also causes a decreased burning rate [42].

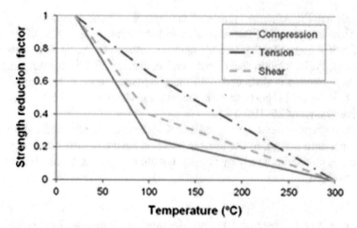

Fig. 2.14 Variation of parallel-to-grain strength elastic modulus of wood with temperature according to EN1995-1-2 [41]

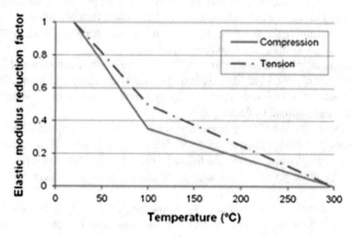

Fig. 2.15 Variation of parallel-to-grain Young's elastic modulus of wood with temperature according to EN1995-1-2 [41]

2.3   Fire Resistance of Wood Structures

The aim is to determine a harmonized procedure for classifying the fire resistance of construction products and building elements. This classification is based on test procedures. The procedure for classifying construction products and building elements is determined on the basis of data from fire resistance tests and smoke tightness tests, which are delimited by the area of the direct application according to the relevant test method [3].

Determination of loading capacity, integrity and insulation are generally required. There are also optional attributes specified, such as radiation, mechanical aspects, self-closing ability, and smoke penetration. The classification based on optional attributes depends on national standards and can be specified individually for certain items under certain conditions.

Loading capacity and stability R is the ability of a structural element to withstand a certain time of fire exposure on one or more sides at a specified mechanical stress without losing its structural strength. The criteria for assessing imminent collapse vary depending on the type of load-bearing element. For elements loaded with bending, such as ceilings and roofs, these criteria include the deformation rate (bending rate) and the limit state for the actual deformation (bending), or for axially loaded elements like columns and walls, criteria are the deformation rate (contraction rate) and the limit state for actual deformation (contraction).

Integrity E is the ability of a structural element having a fire separation function to withstand a fire acting from one side only, without the fire being transmitted to the unstressed side due to the penetration of flames or hot gases. These could cause the ignition of either the unstressed surface or any material in its vicinity.

The assessment of integrity is generally made on the basis of the following three conditions of breach of the integrity criterion during a test:

– cracks or openings exceeding the specified limits;
– ignition of the cotton cushion;
– continuous burning with a flame on the undamaged side.

Integrity must be determined during the test in all three ways, with the cotton pad attached until it ignites, then removed and the test continued until all three conditions have occurred (however, the test owner may stop the test as soon as the desired level has been reached). The time of integrity failure at each condition is recorded. Failure of the load-bearing capacity criterion must also be regarded as a breach of integrity.

The assessment of the integrity of some elements requires additional measurements; for other elements, the integrity need not be determined according to any of the three criteria mentioned in the previous paragraph. The methodology for these cases is determined in specific test standards.

Thermal insulation I is the ability of an element to withstand a fire acting on one side only without the transfer of fire as a result of a significant heat transfer from the stressed side to the unstressed side. The heat transfer must be limited such that neither the stressed side nor any material in its immediate vicinity ignites. The element must also provide a barrier against heat spread sufficient to protect people in its vicinity.

Integrity classification depends whether an element meets the thermal insulation criterion. If it passes the E and I tests, the category is determined by any of the three integrity criteria, whichever was violated first. If an element passes the E test, but not the I test, the value for integrity classification is determined by the time until the integrity criterion is violated by crack or hole occurrence or by the state of continuous flame burning, whichever occurs first.

For all partition elements except doors and shutters, the limit state for determining thermal insulation is an increase in the average temperature on the unstressed surface, limited to 140 °C above the initial average temperature, while the increase in maximum temperature at any point is limited to 180 °C above the initial average temperature.

Radiation-controlled insulation W is the ability of an element to withstand a fire acting on one side only and to reduce the likelihood of fire transfer as a result of significant radiant heat transferred either through an element or from its unstressed side to nearby materials. The element may also be needed to protect people around it. Violation of the integrity criterion by the condition "cracks and openings exceeding the specified limits" or by the condition "continuous flame burning on the unstressed side" automatically means a violation of the radiation criterion. Elements for which radiation is assessed are marked by adding "W" to the classification (e.g. EW, REW). For these elements, the classification is given by the time during which the maximum value of radiation, measured according to the test standard, does not exceed 15 kW/m^2.

Other criteria usually determined are: *Mechanical action* M, *Automatic closing* C, *Smoke tightness* S, *Soot fire resis*tance G, *Protection against fires of coatings* K.

2.4 Eurocode 5

2.4.1 Origin and Development of the Eurocode Standards

The Eurocodes provide general rules for the design of structures and elements for common use, whether of a traditional or innovative nature. Unusual types of structures or special designs are not the subject of Eurocodes and in such cases, designers can request additional expert opinion.

The Eurocode programme includes the following standards, and generally consist of multiple parts:

EN 1990 Eurocode: Basis of structural design.
EN 1991 Eurocode 1: Action on structures.
EN 1992 Eurocode 2: Design of concrete structures.
EN 1993 Eurocode 3: Design of steel structures.
EN 1994 Eurocode 4: Design of composite steel and concrete structures.
EN 1995 Eurocode 5: Design of timber structures.
EN 1996 Eurocode 6: Design of masonry structures.
EN 1997 Eurocode 7: Geotechnical design.
EN 1998 Eurocode 8: Design of structures for earthquake resistance.
EN 1999 Eurocode 9: Design of aluminum structures.

Standard EN 1995-1-2 [43] is intended for customers, for example to formulate their specific requirements, designers, suppliers, and for relevant government bodies. The general objectives of fire protection are to limit the risks in the event of a fire with regard to the individual, society, property, and, where required, the wider environment. The structure must be designed and constructed in such a way that in the event of a fire:

- the load-bearing capacity of the structure is maintained for the required period of fire resistance,
- the occurrence and spread of fire and smoke within the building is prevented,
- the spread of the fire to the surrounding buildings is prevented,
- the persons present can leave the building or otherwise be rescued,
- the safety of rescue teams is taken into account.

The parts of the Eurocodes which lay out requirements for structural design as it pertains to potential fires deal with specific aspects of passive fire protection in terms of the design of structures and parts of structures in terms of adequate load-bearing capacity and limitation of fire spread. The required functions and performance levels can be determined either by an assessment of the nominal (standardized) fire resistance, usually specified in national fire regulations, or by reference to the fire safety engineering assessment of passive and active measures.

There are some additional requirements, which consider, for example:

- installation and maintenance of a stationary fire extinguishing system,
- conditions for persons in the building or relevant fire sector,
- insulation and coating materials, including their maintenance, which are not mentioned in this document, as they are subject to the regulations of the competent authorities.

EN 1995 [10, 43] concerns the design of buildings and civil engineering structures made out of wood (natural, sawn, planed, or in the form of logs, glued laminated timber, or other wood products for load-bearing purposes, such as laminated veneer wood) or made of wood-based materials which are joined together by adhesives or mechanical fasteners. EN 1995 complies with the principles and requirements for the safety and serviceability of structures and with the principles of their design and verification [44]. The EN 1995 standard is intended to be used together with:

EN 1990: 2002 Eurocode—Basis of structural design [44],

EN 1991 Actions on structures [45],

EN 1998 Design of structures for earthquake resistance if the structures are located in seismic areas [46],

EN 1995 [10] only covers the requirements for mechanical resistance, serviceability, durability, and fire resistance of wooden structures. Other requirements, such as thermal or sound insulation, are not included.

Eurocode 5 is divided into the following parts:

EN 1995-1 General: Structural fire design.

EN 1995-1-1 General—Common rules and rules for buildings.

EN 1995-1-2 General—Structural Fire Design.

EN 1995-2 Bridges [47]

EN 1995-2 is related to the general rules in EN 1995-1-1 [10]. The articles in EN 1995-2 supplement the articles in EN 1995-1-1.

EN 1995-1-2 applies to the design of wooden structures in an emergency situation of fire and must be used in conjunction with EN 1995-1-1 [10] and EN 1991-1-2 [43]. It only defines deviations or additions to the design at normal temperatures. EN 1995-1-2 deals only with passive fire protection methods; active fire protection methods are not included in this standard. In addition to the general assumptions of EN 1990: 2002, it is assumed that all passive fire protection systems taken into account in the design will be made accordingly.

EN 1995-1-2 applies to building structures that are required to perform certain functions when exposed to fire, in terms of:

– prevention of premature collapse of the structure (load-bearing function),
– limiting fire spread (flames, hot gases, excessive heat) outside the given area (dividing function)

EN 1995-1-2 sets out the principles and application rules for the design of structures for specified requirements with regard to the mentioned functions and the levels of their fulfillment.

EN 1995-1-2 applies to structures or components which are the subject of EN 1995-1-1 and which are designed in compliance with it.

2.4.2   Related European Standards

Eurocode 5 respects the following standards:

EN300 Oriented strand boards (OSB). Definitions, classification, and specifications [48].

EN 301 Adhesives, phenolic and aminoplastic for load-bearing timber structures; Classification and performance requirements [49].

EN 309 Particleboards. Definition and classification [50].

EN 313 Plywood. Classification and Terminology [51].

EN 314 Plywood. Bonding quality. Requirements [52].

EN 316 Wood fibre boards. Definition, classification, and symbols [53].

EN 520 Gypsum plasterboards. Definitions, requirements, and test methods [54].

EN 912 Timber fasteners—Specifications for timber connectors [55].

EN 1363 Fire resistance tests. General requirements [56].

EN 1365 Fire resistance tests for loadbearing elements: Walls [57].

EN 1365 Fire resistance tests for loadbearing elements: Floors and roofs [58].

EN 1990:2002 Eurocode. Basis of structural design [44].

EN 1991 Eurocode 1: Actions on structures—Part 1–1: General actions—Densities, self-weight, imposed loads [43].

EN 1993 Eurocode 3. Design of steel structures—Part 1–2: General rules—Structural fire design [46].

EN 1995 Eurocode 5: Design of timber structures—Part 1–2: General—Structural fire design [44].

EN 12,369 Wood-based Panels—Characteristic Value for Structural Design— OSB, particleboards, and fibreboards [59].

EN 13,162 Thermal insulation products for buildings [60].

ENV 13,381–7 Test methods for determining contribution to the fire resistance of structural members. Applied protection to timber members [61].

EN 13,986 Wood-based panels for use in construction—Characteristics, evaluation of conformity and marking [62].

EN 14,081-1Timber structures. Strength graded structural timber with rectangular cross section. General requirements [63].

EN 14,080 Timber structures. Glued laminated timber and glued solid timber. Requirements [64].

EN 14,374 Timber structures. Structural laminated veneer lumber. Requirements [65].

The Eurocodes recognize the responsibility of the national authorities of each Member State and preserve their right to set at national level values relating to safety, which may vary from one Member State to another.

2.4.3 Design Principles

If mechanical resistance is required in the event of a fire, the structures must be designed and constructed in such a way as to retain their load-bearing function during the relevant fire exposure. Where division into fire compartments is required, the elements forming the boundary of the fire compartment, including joints, must be designed and constructed so as to retain their partitioning function during the fire exposure. It is crucial that: there is no integrity failure; there is no insulation ability failure; thermal radiation on the non-exposed side is limited. If the temperature on the non-exposed side is lower than 300 °C, fire risk due to thermal radiation is not present [44].

The deformation criterion must be used if the protective means or design criteria for fire dividing structures require the deformation of the supporting structure to be taken into account. Deformation of the load-bearing structure need not be taken into account in the following cases:

- the effectiveness of fire protection arrangements has been proven,
- fire dividing walls satisfy the requirements of nominal fire exposure,

Eurocode 5 differs between two forms of fire exposure—nominal and parametric.

For a nominal exposure, structural elements must satisfy criteria R, E and I as follows:

- dividing function only: integrity E, and where required, insulating ability I
- load bearing ability only: mechanical resistance R
- dividing function and load bearing ability: R and E criteria, and I criterion if required.

The criterion "R" is presumed to be fulfilled if the load-bearing function is maintained during the required fire exposure time.

The criterion "I" is presumed to be met if after the increase in average temperature over the whole surface which is not exposed to fire, the temperature is not more than 140 °C and the maximum temperature at any point on that surface does not exceed 180 °C.

When exposed to a parametric fire, the load-bearing function must be maintained throughout the fire, including the cooling phase, or for the required period of time, provided that the normal temperature is 20 °C, and the following shall be used to verify the division function:

- the average temperature of the non-exposed surface should not be higher than 140 °C and maximum temperature at any point on that surface not higher than 180 °C during the heating stage until the maximum temperature of the gases is reached
- the average temperature of the non-exposed surface should not be higher than $\Delta\theta 1$ (200 °C) and the maximum temperature at any point on that surface not higher than $\Delta\theta 2$ (240 °C) during the cooling phase.

Design material values and resistances are described in the standard in detail. Verification methods allow the analysis of structural elements, structural parts, and the entire structure. It must also be noted that Eurocode 5 enables fire resistance calculation only up to 60 min.

2.4.4 Examples

In this subchapter, we will give several examples of fire resistance calculated on the basis of the Eurocode 5 calculation, separately for beams and for structural wall parts (board sheeting materials). For beams, we present only the criterion R [28] which are given in the Tables 2.1 and 2.2.

There are four types of wall materials displayed in 4 graphs, in relation to their thickness. Figure 2.16 shows wood-based boards (SM-Spruce), Fig. 2.17 shows OSB boards, Fig. 2.18 shows A and H type plasterboards, and Fig. 2.19 shows F-type plasterboards.

Table 2.1 Fire resistance (minutes) of a glued wooden beam, 120 × 80 mm cross-section, mechanically stressed by compression, tension and bending, exposed to fire from one, two, three and four sides

Compression	Tension	Bending	Fire exposure
50	60	60	
10	37	28	
9	35	29	

Table 2.2 Fire resistance (minutes) of a glued wooden beam, 180×100 mm cross-section, mechanically stressed by compression, tension and bending, exposed to fire from one, two, three and four sides

Compression	Tension	Bending	Fire exposure
60	60	60	
28	50	43	
27	50	42	

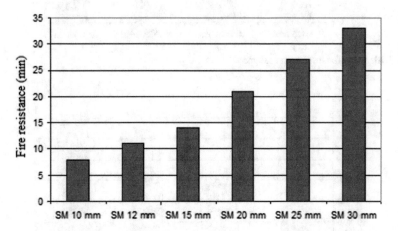

Fig. 2.16 Wood-based boards (Spruce)

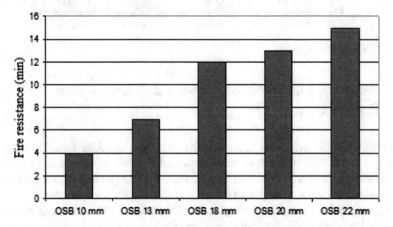

Fig. 2.17 OSB boards

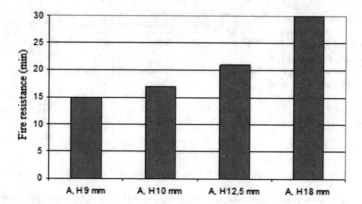

Fig. 2.18 Plasterboards, A and H type

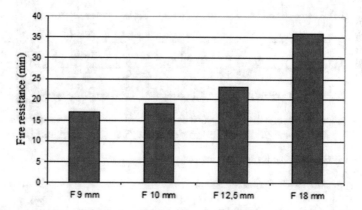

Fig. 2.19 Plasterboards, F-type

2.5   Evaluation of Wood for Wooden Buildings—Fire Protection

This chapter in no way aims to summarize all of the scientific knowledge which directly—to a greater or lesser extent—leads to the fire safety of wooden buildings. The improvement of scientific instrumental methods that can investigate the properties of wood at elevated temperatures, as well as the development of new methods of computer technology, have brought about the improvement and enhancement of scientific knowledge that is directly applicable in practice.

The process of thermal degradation is very complex. Its analysis must begin with the thermal degradation of the basic building elements of wood: hemicelluloses, cellulose, and lignin. Clarification of the thermal degradation of individual building components clarified the whole view of the thermal degradation of wood as a building element in a fire. In addition to chemical composition, the physical properties of wood and wood-based materials applied in wooden buildings also play an important role. Among many characteristics, density and thickness of the material should be highlighted, as they have a direct effect on the combustion process, heat transport, and the development of fire.

Nearly 40 years of effort by scientists from around the world have provided a new perspective on changes in the mechanical properties of wood at elevated temperatures. This research was complicated by the conditions of individual experiments and interpretation results, not to mention the nature of wood as a heterogeneous organic material with a variable moisture content. The work has led to reliable values of relative change in the mechanical properties of wood (for bending, compression, tension, and shear). These values can be then implemented into the creation of national and international standards and other regulations.

A normative rule-making process was mentioned, as it has changed in terms of the testing methods for wood and wood composites in relation to fire. The conditions described in Sect. 2.3 are particularly related to the fire resistance of building structures. A new concept of "reaction to fire" (or combustibility) has been introduced, which is determined under completely new conditions and using new testing procedures.

The Eurocode standards are also part of the new evaluation procedure. They allow fire resistance calculation of timber structures based on the analyzed knowledge, whether for single elements, structural parts, or the entire structure. Eurocode 5 will make it possible to assess the behavior of a load-bearing element, in particular its charring. The creation of a charred layer, which ensures the auto-retarding character of the wood-burning process, has reached legislative outputs, which helps to more objectively assess the behavior of wood in the event of a fire. The Eurocode procedure also includes the influence of auxiliary items such as metal fasteners, adhesives, and secondary materials—filling insulating materials, sheeting material, and all other materials which are part of timber structures. Such a calculation can substitute relatively expensive structural tests in certified facilities.

In Chap. 3, we will address the concept of large-scale tests. After all that has been presented in this chapter—new scientific knowledge, legislation, quality testing of materials' reaction to fire, and testing of structures' fire resistance—it is vital to consider experiments in the form of large-scale tests. Only a large-scale test can reveal the actual behavior of a wooden building in conditions of a real fire. Therefore we decided, with the help of our partners, to perform such a large-scale test of a two-story wooden building—in the hope, of course, that this building, with its quality design craftsmanship that take into account the relevant fire-resistance principles, would withstand the fire.

References

1. Osvald A, Balog K (2017) Horenie dreva. Vydavateľstvo Technickej univerzity vo Zvolene, Zvolen, p 106. ISBN 978–80–228–2953–3
2. Ševeček P, Netopilová M (1992) Nauka o materiálu. (Material science). VŠB TU v Ostrave, Ostrava pp 184
3. Osvald A (1995) Drevostavba ≠ požiar. (Wooden building ≠ fire). Technická univerzita vo Zvolene, Zvolen, p 336. ISBN 978–80–228–2220–6
4. Schaffer EL (1973) Elevated temperature effect on the longitudinal mechanical properties of wood. PhD Thesis, Department of Mechanical Engineering, University of Wisconsin, Madison, WI 1970
5. Gerhards CHC (1982) Effect of moisture content and temperature on the me-chanical properties of wood: an analysis of immediate effects. Wood Fiber 14(1):4–36. ISSN: 07356161
6. Bučko J, Klaudová A, Kačík F (1994) Vacuumtrockung des Laub- und Nadelholz. Theorie und Praxis der Vacuum-Schnittholztrocknung. Internationales Wissenschaftliches Symposium. Technical University in Zvolen, Zvolen, pp 96–104
7. Fengel D, Wegener, G (1989) Wood. Chemistry, ultrastructure, reactions. Walter de Gruyter, Berlin, Germany, pp 26–226
8. Kačíková D (2004) Vplyv nízkoteplotnej degradácie na zmeny vybraných che-mických a mechanických charakteristík smrekového dreva. (Effect of low-temperature degradation on changes in selected chemical and mechanical characteristics of spruce wood.) Zborník Wood and Fire Safety. 2004. Zvolen: Technická univerzita vo Zvolene. Not numbered. ISBN 80–228–1321–4
9. Karlsson B, Quintiere, JG (2000) Enclosure fire dynamics. CRC Press, London. ISBN 0-8493-1300-7
10. 1995-1-1 Eurocode 5: Bemessung und Konstruktion von Holzbauten. Teil 1–1: Allgemeines. Allgemeine Regeln und Regeln fűr den Hochbau
11. Okuyama T (1974) Effect of strain rate on mechanical properties of wood. IV. On the influence of the rate of deflection and the temperature to bending strength of wood. J Jap Wood Res Soc 20(5):210–16. ISSN 1435–0211
12. Požgaj A et al (1997) Štruktúra a vlastnosti dreva. (Wood structure and properties), no 2. Príroda, Bratislava, p 488. ISBN 80–07–00960–4
13. Green DW et al (1999) Adjusting modulus of elasticity of lumber for changes in temperature. Wood Eng 10:82–94
14. Glos P, Henrici D (1991) Biegefestigkeit und Biege-E-Modul von Fichtenbau-holz im Temperaturbereich bis 150 °C. Holz Roh-und Werkstoff 49:417–22. ISSN 0018–3768
15. Comben AJ (1964) The effect of low temperatures on the strength and elastic properties of timber. J Inst Wood Sci 13:44–55. ISSN 0043–7719
16. Partl M, Strassler H (1977) Effect of temperature on the static and impact bending behavior of spruce wood. Holzforsch. Holzwerwert 29(5):94–101. ISSN 0018–3830
17. Sulzberger PH (1953) The effect of temperature on the strength of wood, plywood and glue joints. Aeronaut. Res. Consultative Com. Rep. ACA–46. Melbourne, Australia
18. Tsuzuki K, Takemura T, Asano I (1976) Physical properties of wood-based materials at low temperatures I. The bending strength of wood as related to temperature and specific gravity. J Jap Wood Res Soc 22(7):381–86. ISSN 1435–0211
19. Östman BL (1985) Wood tensile strength of temperatures and moisture content simulating fire conditions. Wood Sci Technol 19(2):103–16. ISSN: 0043 7719
20. Östman BL (2010) Fire safety in timber buildings. Technical guideline for Europe. SP report 2010, vol 19. ISBN 978–91–86319–60–1
21. Nyman C (1980) The effect of temperature and moisture on the strength of wood and glue joists. Technical Research Centre of Finland (VTT). Report Forest Products Lab, p 6
22. Knudsen RM, Schniewind AP (1975) Performance of structural woodmembers exposed to fire. Forest Prod J 25(2):23–32. ISSN:00157473

23. Mäger KN, Tiso M, Just A (2020) Fire design model for timber frame assemblies with rectangular and i-shaped members. In: Makovicka Osvaldova L et al (eds) WFS 2020, Wood & fire safety. © Springer Nature Switzerland AG 2020, pp 268–274. https://doi.org/10.1007/978-3-030-41235-7_40

24. Schaffer EL (197) State of structural timber fire endurance. Wood Fiber 9(2):145–170. ISSN: 07356161

25. Pozdieiev S et al (2020) Research of wooden bearing structures behavior under fire condition with use advanced methods of fire resistance calculation considering eurocode 5 recommendation. In Makovicka Osvaldova L et al (eds) WFS 2020, Wood & fire safety. © Springer Nature Switzerland AG 2020, pp 326–332. https://doi.org/10.1007/978-3-030-41235-7_48

26. Regináč L (1990) Náuka o dreve II. (Wood science II). Technická univerzita, Zvolen p 424. ISBN 80–228–0062–7

27. Osvald A, Balog K (1990) Zmeny vo vlastnostiach smrekového dreva po tepelnom namáhaní: I. Mechanické a fyzikálne vlastnosti. (Changes in the properties of spruce wood after heat stress: I. Mechanical and physical properties. Drevo 45:103–5. ISSN 0012-6144

28. Osvald A (1995) Vplyv vyšších teplôt na tlakovú pevnosť smrekového dreva. (The impact of higher temperatures on the compressive strength of spruce wood). Zborník vedeckých prác DF VŠLD Zvolen, pp 285–296. ISBN 80–05–00578–4

29. Kollman F (1940) The mechanical properties of wood of different moisture contents in –200 °C to +200 °C temperature range. VDI-Forschungsh 403(11):1–18

30. Kollman F (1960) The dependence of the elastic properties of wood on temperature. Holz Roh-Werkst 18(8):304–314. ISSN 0018- 3768

31. Okuyama T (1975) Effect of strain rate on mechanical properties of wood. V. On the influence of temperature on bending strength. J Jap Wood Res Soc 20(5):210–16. ISSN 1435–0211

32. Sellevold E J et al (1975) Low temperature internal friction and dynamic modulus for beech wood. Wood Fiber 7(3):162–169. ISSN: 07356161

33. Kitahara RN, Matsumoto T (1974) Temperature dependence of dynamic mechanical loss of wood. J Jap Wood Res Soc 20(8):349–54. ISSN 1435–0211

34. Sadoh T (1981) Viscoelastic properties of wood in swelling systems. Wood Sci Tech 15(1):57–66. ISSN 0043-7719

35. Young RA (1978) Thermal transitions of wood polymers by torsional pendulum analysis. Wood Sci 11(2):79–101. ISSN 0043–7719

36. Law KN (1981) Koran Z (1981) Torsional-shear stress of wood at varioutemperatures. Wood Sci Technol 15(3):227–235

37. Stone JE (1965) The rheology of cooked wood. II. Effect of temperature. Tappi 38(8):452–55

38. Martinka J et al (2016) Investigation of the influence of spruce and oak wood heat treatment upon heat release rate and propensity for fire propagation in the flashover phase. Acta Facultatis Xylologiae Zvolen 58(1):5–14. https://doi.org/10.17423/afx.2016.58.1.01. ISSN: 1336–3824

39. Chovanec D, Osvald A (1992) Spruce wood structure changes caused by flame and radiant source. Technical university in Zvolen, Zvolen 62. ISBN 80–228–0182–8

40. Iringova A (2017) Lightweight building envelopes in prefabricated buildings in terms of fire resistance. In: MATEC Web of conferences, vol 117, art no 00062. ISSN: 2261 236X 4

41. Maraveas C, Miamis K, Matthaiou CE (2015) Performance of timber connections exposed to fire: a review. Fire Technol 51:1401–1432. https://doi.org/10.1007/s10694-013-0369-y

42. Tran HC, White RH (1992) Burning rate of solid wood measured in a heat release rate calorimeter. Fire Mater 16:197–206. ISSN: 0308–0501

43. EN 1995-1–2 Eurocode 5: Design of timber structuresPart1–2: General Structural fire design

44. EN 1990 (2002) Eurocode. Basis of structural design

45. EN 1991 (2002) Eurocode 1: Action on structures

46. EN (1998) Eurocode 8: Design of structures for earthquake resistance

47. EN 1995-2 Eurocode 5: Bridges

48. STN EN 300 (1999) Oriented Strand Boards (OSB). Definitions, classification and specifications

49. EN 301 Adhesives, phenolic and aminoplastic, for load-bearing timber structures-Classification and performance requirements.
50. EN 309 (1992) Particleboards. Definition and classification
51. EN 313-1 (2001) Plywood. Classification and terminology. Part 1: Classification
52. EN 314-1 (2005) Plywood-Bonding quality-Part 1: Test methods
53. EN 316 (2009) Wood fibre boards. Definition, classification and symbols
54. EN 520 (2004) Gipsplatten. Definitionen, Anforderungen, Prüfverfahren
55. EN 912 (2001) Timber fasteners-specifications for connectors for timbers
56. EN 1363 (1999) Fire resistance tests
57. BS EN 1365-1 (1999) Fire resistance tests for non-loadbearing elements. Part 1: Walls
58. EN 1365-2 (2000) Fire resistance tests for loadbearing elements. Part 2: Floors and roofs
59. EN 12369–1 (2001) Wood-based panels. Characteristic values for structural design. Part 1: OSB, particleboards and fibreboards
60. STN EN 13162 (2003) Thermal insulation products for buildings. Factory made mineral wool (MW) products. Specification
61. ENV 13381–7 (2019) Test methods for determining the contribution to the fire resistance of structural members. Part 7: applied protection to timber members
62. EN 13986:2004 Wood-based panels for use in construction. Characteristics, evaluation of conformity and marking
63. STN EN 14081–1 (2005) Drevené konštrukcie. Timber structures-strength graded structural timber with rectangular cross section-Part 1: General requirements
64. STN EN 14080 (2005) Drevené konštrukcie. Timber structures. Glued laminated timber and glued solid timber. Requirements
65. EN 14374 (2004) Timber structures. Structural laminated veneer lumber. Requirements
66. England P, Iskra B (2020) Australian building code change-eight-storey timber buildings. In: Makovicka Osvaldova L et al (eds) WFS 2020, Wood & Fire Safety. © Springer Nature Switzerland AG 2020, pp 219–225. https://doi.org/10.1007/978-3-030-41235-7_33
67. Iringova A (2017) Revitalisation of external walls in listed buildings in the context of fire protection. In: Procedia engineering, vol 195, pp 163–170. ISSN: 1877–7058, 2017
68. Kasymov D et al (2020) Thermography of wood-base panels during fire tests in laboratory and field conditions. In: Makovicka Osvaldova L et al (eds) WFS 2020, Wood & fire safety. © Springer Nature Switzerland AG 2020, pp 203–209. https://doi.org/10.1007/978-3-030-41235-7_31
69. Niemz P (1993) Physik des Holyes und der Holzwerkstoffe. DRW–Verlag, Dresden, p 243
70. Sandanus J, Sógel K (2011) Timber structures–exercises and calculations, University script, Chapter No.7: Design of connections in a timber hall structure (in Slovak), pp 77–101
71. EN 13501-2 (2004) Fire classification of construction products and building elements. Part 2: Classification using data from fire resistance tests, excluding ventilation services

Chapter 3
Model Fire in a Two-Story Wooden Building

Abstract The chapter presents the necessity of and reasoning behind large-scale fire tests. It describes in detail the construction of a model two-story wood-based building which was subjected to fire in such a test. The chapter contains basic technical documentation of the wooden building, partial photo documentation from the construction, and photos from the course of the fire. The temperature was graphically recorded in each location in the building and in individual elements of its construction.

Keywords Two-story wooden building · Model fire · Fire resistance

3.1 Basic Idea of Model Tests

Large-scale tests can also include standardized tests for fire resistance. In addition to existing standardized tests, other large-scale tests are sometimes performed. These tests do not have a legal character but provide important information on the behavior of the materials in the structure as well as the structure itself in the event of a fire, as observed in the simulation.

Large-scale tests are performed not only in wooden constructions, but also in buildings with load-bearing steel structures, new structural elements and solutions, thermal insulation systems, etc. These tests are very costly, but they are still needed and therefore used.

A building that undergoes a fire test should be fitted with measuring equipment that helps to detect weak spots in the structure or material in conjunction with other material, the quality of joints, the quality of craftsmanship, or the impact of fire treatment and other modifications. Proper installation of measuring equipment, thermocouples, airflow meters, or other measuring devices is very important for the test; this includes the choice of location in the building to be tested, as well as the correct choice and scope of measuring equipment.

The location, size, and extent of the fire are equally important. Only a model fire that simulates real fire conditions will help generate accurate data with a high technical or scientific value. As can be expected, preparing a model fire test is not easy and requires extensive preparation, not only in terms of construction of the

Fig. 3.1 Preparation of a six-story wooden structure for a test fire, England [1]

tested building but also the preparation of the measurement and test itself. Despite all these challenges and high financial costs, we were able to prepare and implement such a test with the cooperation of the participating companies.

We are not the only researchers performing such model tests as documented in the attached photos. Figure 3.1 shows the preparation of a model test of a six-story wooden building in England, and Fig. 3.2 depicts model fires along the facade of a building in Germany.

3.2   Structural Fire Test Scheme of a Two-Story Building

The model fire experiment of a two-story building was coordinated by the Association of Wood Processors in cooperation with partner companies, the Technical University in Zvolen and the University of Žilina. This demonstration took place on the premises of the certified testing laboratory FIRES s.r.o in Batizovce during during 7th International Scientific Conference Wood & Fire Safety 2012. FIRES s.r.o. is one of the youngest European testing and certification institutions which is authorized to issue certificates of quality and correctness of construction procedures in terms of fire resistance. This institution also provided us with a measuring apparatus with which the results presented in this publication were obtained.

Fig. 3.2 Building after facade fire test Germany [1]

The two-story wood-based building was subjected to a test simulating the conditions of a real fire. The building was made of a prefabricated panel structure of 4.9 × 3.7 m, with a height of 5.6 m and a load-bearing wooden frame filled with mineral wool. The perimeter walls were supplemented with an installation layer with a mineral wool filling and sheathed with a contact thermal insulation system—two mineral wood-based walls and the other two walls made of wood-fiber boards.

The structure was not a standard wooden building, as the goal of the experiment was to test several materials. All materials used were applied in such a way to correspond with Eurocode 5. On the basis of this calculation, the thickness, density, and other physical parameters of the material were determined. Therefore, each wall represented a different material composition. The inner surfaces were lined with plasterboards. The compositions of the perimeter walls, ceiling, roof, and partition walls were designed with regard to the fire resistance of the building system for low-energy and energy-passive multi-story wood-based houses, while they were dimensioned for fire resistance in three limit states (R—resistance, E—integrity and I—thermal insulation) for 45 min. The building had windows with standard fire resistance, which were placed above one another on each floor. The effect of a fire curtain, to prevent the spread of fire along the facade to the upper parts of the building, was also tested.

The overall project of the building, dimensions, material composition of individual walls, partition walls, and ceilings, as well as details of construction solutions, are

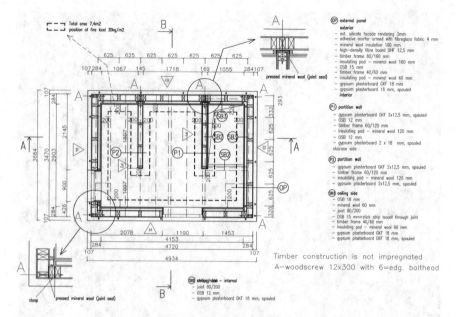

Fig. 3.3 Floor plan of the ground floor of the model two-story building

shown in Figs. 3.3, 3.4, 3.5, 3.6 and 3.7. The joint between two wooden elements was insulated using compressed mineral wool—see Detail A (Fig. 3.3). This modification was made in order to prevent leakage in the event of fire, causing a faster air flow and thus overheating of the opening, which could cause the fire to enter the cavities of the building and endanger the load-bearing parts of the structure.

3.3 Construction of a Two-Story Building

The construction process is documented in Figs. 3.8, 3.9, 3.10, 3.11, 3.12, 3.13, 3.14, 3.15, 3.16, 3.17, 3.18, 3.19, 3.20, 3.21, 3.22, 3.23, 3.24, 3.25, 3.26 and 3.27, which were selected from more than 100 photographs taken during the process in order to show the details and the overall design of the model building. Figure 3.8 shows the foundations, Figs. 3.9, 3.10, 3.11, 3.12 and 3.13 present the erecting of walls and structures of the ground floor, Fig. 3.14 displays ground floor ceiling installation, and Fig. 3.15 illustrates the erection of the north external wall on the first floor. Installation of the roof trusses is shown in Fig. 3.16. Various details (roof, interior) are shown in Figs. 3.17, 3.18, 3.19, 3.20, 3.21, 3.22, and preparation works prior to exterior thermal insulation installation are shown in Fig. 3.23.

Figures 3.24, 3.25, 3.26 and 3.27 display the finishing of exterior details and also selected details of fire protection measures. There is an example of an external wall insulation system shown in Fig. 3.25, and final interior finish in Fig. 3.27.

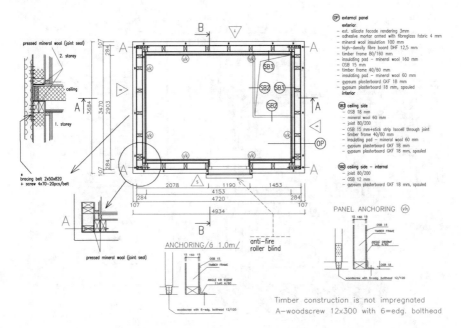

Fig. 3.4 Floor plan of the first floor of the model two-story building

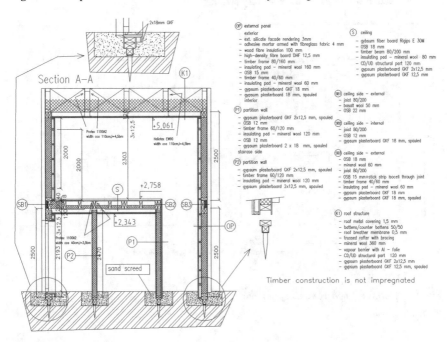

Fig. 3.5 Section A-A of the model two-story building

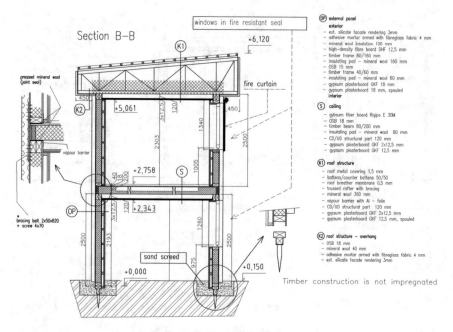

Fig. 3.6 Section B-B of the model two-story building

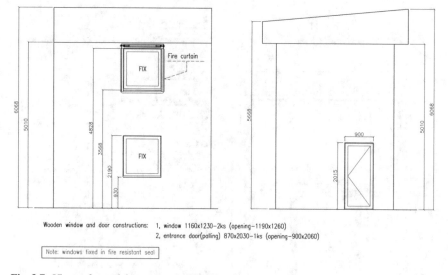

Wooden window and door constructions: 1, window 1160x1230–2ks (opening–1190x1260)
2, entrance door(paling) 870x2030–1ks (opening–900x2060)

Note: windows fixed in fire resistant seal

Fig. 3.7 Views of a model two-story building

Fig. 3.8   Foundation of the building by means of ground screw anchors

Fig. 3.9   North wall at the ground floor level

Fig. 3.10 Prefabricated panel fixed to horizontal wooden sole plate

Fig. 3.11 Second wall installation of the ground floor

Fig. 3.12   Wall joint detail of the ground floor

Fig. 3.13   Wall installation at ground floor level

Fig. 3.14 Ceiling panel installation

Fig. 3.15 North wall installation at the first floor level

Fig. 3.16 Installation of roof trusses

Fig. 3.17 Detail of a roof truss, covered with fire resistant fabric

3.4 The Test Structure Fire—Photos

Selected pictures of the modeled structural fire and its details are shown in Figs. 3.28, 3.29, 3.30, 3.31, 3.32, 3.33, 3.34, 3.35, 3.36, 3.37, 3.38, 3.39, 3.40, 3.41, 3.42, 3.43, 3.44, 3.45, 3.46 and 3.47.

Fig. 3.18 Thermal insulation in the roof

Fig. 3.19 Ceiling construction details, with wood elements wrapped in fire-resistant fabric

Fig. 3.20 Entrance to the first floor (stair void), thermal insulation in front of the main timber frame (space for plumbing and wiring installations)

Fig. 3.21 Plasterboards installation, different types of thermal insulation application

Fig. 3.22   Plasterboard installation, thermal bridges protection by the fire-resistant fabric

Fig. 3.23   Scaffolding for external insulation installation

Fig. 3.24 Thermocouples on exterior side of timber frame, prior to the external insulation (position "Y," see Fig. 3.49)

Fig. 3.25 External wood fibre insulation. Thermocouple (Y-type) is visible, placed between the wall layers

Fig. 3.26 Finished exterior—lower window

Fig. 3.27 A window from the interior. Final finish of plasterboards (without a paint coat)

Fig. 3.28 The two-story building

Fig. 3.29 Position of thermocouples on the exterior surface

Fig. 3.30 Position of thermocouples on the interior surface

Fig. 3.31 Connection of thermocouples to the central unit

Fig. 3.32 Fuel layout in interior. The fuel was distributed evenly on the ground floor, to achieve the value of 60–70 kg/m^2

Fig. 3.33 Fuel arrangement composed of OSB and air-dried spruce timber. 10 l of diesel was used to initiate the fire, located in the tin container

Fig. 3.34 Fuel arrangement composed of OSB and spruce timber (50 kg/m^2)

Fig. 3.35 The fire 3 min after the fuel ignition. The black smoke is caused by the diesel fuel burning

Fig. 3.36 Fully developed fire in the interior, 30 min after ignition (see Fig. 3.68 for temperature development)

Fig. 3.37 The fire in its 38th minute

Fig. 3.38 Windows and door after the fire

Fig. 3.39 Interior after the fire. Plasterboards collapse in the area of the fire ignition (space of the extensive fire)

Fig. 3.40 The floor, burned fuel, and plasterboards damaged at the entrance to the building on the ground floor level

Fig. 3.41 The ceiling in the space of extensive fire, after plasterboard removal (thermocouple C1)

Fig. 3.42 Window view after the fire

Fig. 3.43 Interior window opening (sill) after the fire

Fig. 3.44   Exterior detail of the window, protected by the fire retardant fabric. Fire protective barrier Stöbich

Fig. 3.45   Influence of the metal fasteners on wood composite carbonisation, as a result of their high thermal conductivity

Fig. 3.46 Influence of the metal fasteners on wood element carbonisation, as a result of their high thermal conductivity

Fig. 3.47 Intact timber frame structure, after sheathing board material removal

3.5 Assessment of the Fire

The reference sample building resisted the fire exposure well, as documented by photos in the previous chapter. It was proved that all the structural parts and building elements can in fact withstand real fire conditions, as they are subject to fire resistance certification. But, the experiment cannot be evaluated only visually; the monitored data of test conditions need to be processed and interpreted as well. The position of the thermocouples and their locations in the building are shown in Figs. 3.48 and 3.49.

Fragments of these pictures (Figs. 3.50 and 3.53) show details to be assessed and the evaluation is documented by temperature over time recorded by the relevant thermocouple. Each floor (ground and first floor) and similarly each wall (East, North, West, and South) was evaluated independently. The color designation of the individual positions (thermocouples) corresponds to the color of the curve, which represents the measured temperature by the given thermocouple.

The overall assessment is depicted in Figs. 3.50, 3.51, 3.52, 3.53, 3.54, 3.55 and 3.56 for the ground floor, in Figs. 3.57, 3.58, 3.59 and 3.60 for the first floor, Figs. 3.61, 3.62, 3.63, 3.64, 3.65, 3.66 and 3.67 for the East wall, Figs. 3.68, 3.69, 3.70, 3.71, 3.72, 3.73, 3.74 and 3.75 for the West wall, Figs. 3.76, 3.77 and 3.78 for the North wall, and Figs. 3.79, 3.80, 3.81, 3.82 and 3.83 for the South wall.

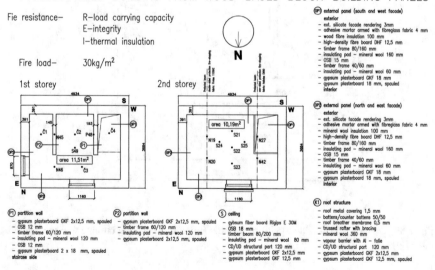

Fig. 3.48 Basic scheme—material assemblies and thermocouples location

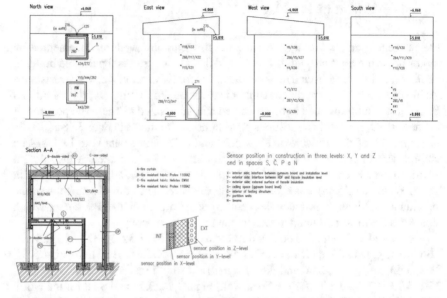

Fig. 3.49 Basic scheme—side elevations and section A-A. Location of thermocouples

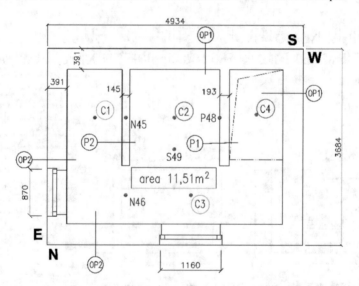

Fig. 3.50 Position scheme (floor plan) of thermocouples for measuring burning fuel temperature (C1 - C4)

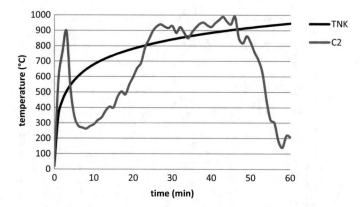

Fig. 3.51 Normalized temperature curve (black) and temperature of C2 thermocouple (red) placed above diesel container

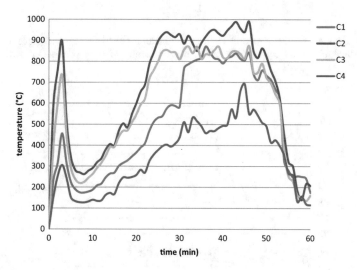

Fig. 3.52 Temperature curves for C1 to C4 thermocouples

3.5.1 Fire Assessment—Ground Floor

These thermocouples were placed in the area under the ceiling. They measured the temperature of the burning fuel (fire) that heated the entire space as well as the temperature of the building structure (walls and ceilings). The color-coding of the thermocouples corresponds to the coloring of the temperature curves in the following figures. The fuel was distributed under all thermocouples (see Figs. 3.32, 3.33, and 3.34).

Figures 3.51 and 3.52 show the temperature curves of individual thermocouples. Thermocouple C2 was in a relatively enclosed space limited by a wall, ceiling, and two partition walls, which extended to about two-thirds of the width of the building.

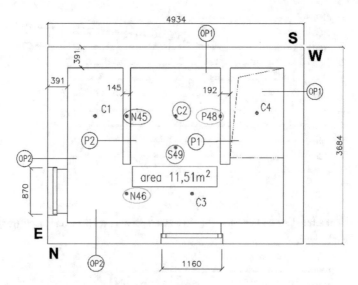

Fig. 3.53 Position scheme of thermocouples (floor plan): N45 and N46 thermocouples on beams, S49 thermocouple on the ceiling (covered by plasterboard), P48 thermocouple on partition wall (covered by plasterboard)

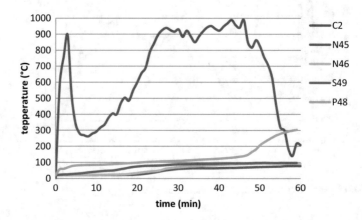

Fig. 3.54 Temperature curves for thermocouples C2, N45, N46, S49, P4

In this space, the fuel was ignited by an initiator, which consisted of 10 L of diesel and OSB boards along with spruce timber. The diesel was poured into a shallow tin container of about 1 m². For this reason, we refer to this place as the place of fire and thermocouple C2 as the reference point from the place of fire.

Figure 3.51 shows the course of the temperature of thermocouple C2 and the temperature standard curve for determining the fire resistance of an internal fire. As can be seen from the course of thermocouple C2, the temperature slightly exceeds the standard temperature curve from the 20th to the 50th minute.

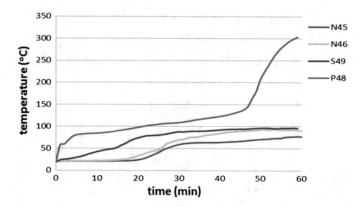

Fig. 3.55 Temperature curves for thermocouples N45, N46, S49, P48

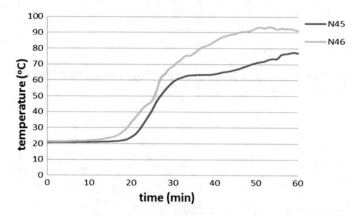

Fig. 3.56 Temperature curves for thermocouples N45 and N46

The temperature peak around the second minute was caused by the ignition and flame of the burning diesel, and the subsequent rapid heating of the thermocouple, which was directly above the container. After ignition of the solid fuel, the temperature gradually rose until it reached the required values, in our case up to the values of the standardized temperature curve. Based on the above, we can conclude that the experiment had corresponding thermal stress in the construction of a two-story wood-based building.

The whole course of thermal stress does not completely coincide with the standard temperature curve, which is to be expected, as it was not a standard laboratory experiment, regulated by fuel from burners with strictly regulated outputs. It can be stated that the conditions in the model test were closer to the conditions of an actual fire.

Thermocouple C3 was located above another fuel source which ignited later, and was in the path of airflow between the door and the upper floor in a horizontal

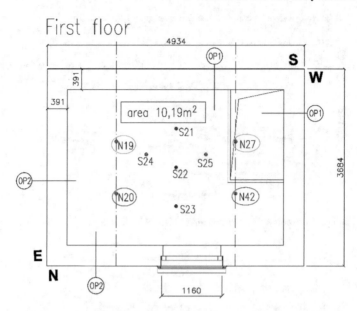

Fig. 3.57 Upper floor level (floor plan)—position of thermocouples N19, N20, N27 and N42 thermocouples on roof trusses

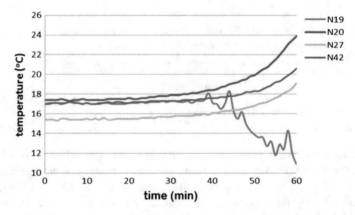

Fig. 3.58 Temperature curves for thermocouples N19, N20, N27 and N42 attached to roof trusses

plane. Thermocouple C1 initially measured lower temperatures, as the fuel had to gradually ignite and start burning outside the ideal airflow, which caused a delay in temperature rise, but from about the 30th minute it measured the same temperatures as thermocouple C3. This means that from this time, the fire spread evenly throughout the first floor. The fuel that was used there was in greater quantities in order to represent an accidental fire load, as it would be in a standard interior fire.

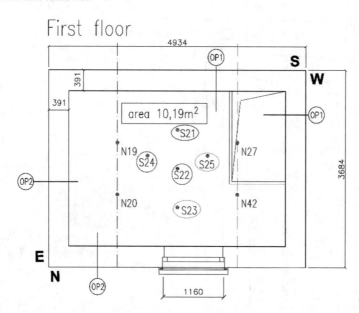

Fig. 3.59 The upper floor level—position scheme of ceiling thermocouples (floor plan) S21, S22, S23, S24 and S25 (see sect. A-A, Fig. 3.40)

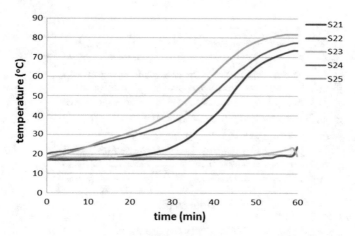

Fig. 3.60 Temperature curves for S21 to S25 thermocouples

Thermocouple C4 measured the lowest temperatures, as it was placed in a space with a strong horizontal and vertical airflow between the door and the second floor of the building. The course of the curve copies the development of temperatures in the room, but the thermocouple cooled down by the air stream (ambient temperature

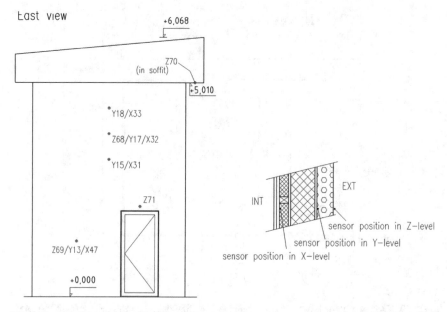

Fig. 3.61 East elevation of the tested building with thermocouple layout and detail of wall assembly with three different types of thermocouple locations within the wall—plane X, Y and Z

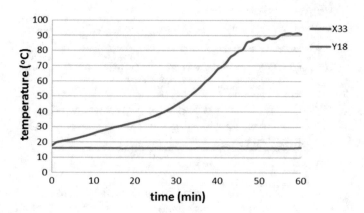

Fig. 3.62 Temperature curves for X33 and Y18 thermocouples

13.5 °C) measured the lowest temperatures. The temperature curves of all thermocouples that checked the fuel and were in the space under the ceiling are shown in Fig. 3.52.

Figure 3.53 shows the location of thermocouples that measured the temperature of the space C2, the partition walls P48, the beams N45 and N46, and the ceiling S49.

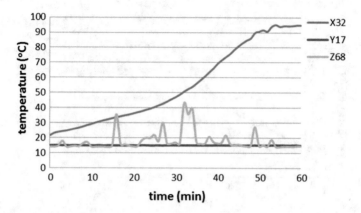

Fig. 3.63 Temperature curves for X32, Y17 and Z6 thermocouples

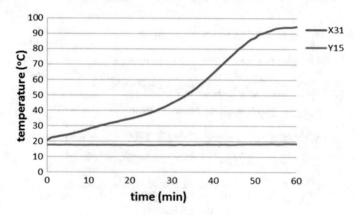

Fig. 3.64 Temperature curves for X31 and Y15 thermocouples

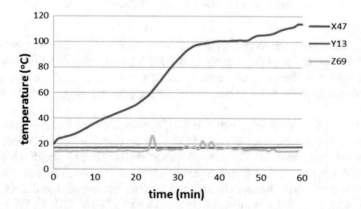

Fig. 3.65 Temperature curves for X47, Y13 and Z69 thermocouples

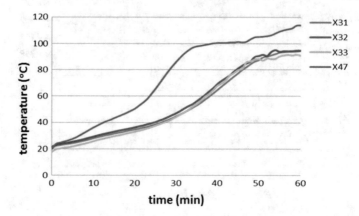

Fig. 3.66 Temperature curves for X31, X32, X33 and X47 thermocouples

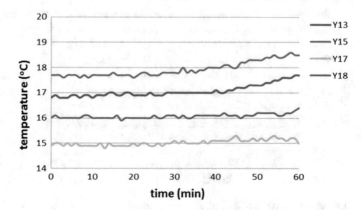

Fig. 3.67 Temperature curves for Y13, Y15, Y17 and Y18 thermocouples

Figure 3.54 shows the course of temperatures measured by individual thermocouples. Of course, the temperature range is at 1000 °C, as thermocouple C4 measured the temperature in the space above the flame. The color-coding of the thermocouples corresponds to the color adjustment of the temperature curves in the following figures.

If we omit the measurement of this temperature from the same measurement (thermocouple C2) we can see in Fig. 3.55 that the maximum temperature reached in the room was just above 300 °C, recorded with thermocouple P48, which was behind the plasterboard partitions in the room where the fuel was ignited. The temperature of 100 °C was exceeded in the 20th minute and the temperature increase did not occur until the 45th minute. Until then, the temperature was below 150 °C. We assume that it was at this time that the thermal insulation was weakened by the damaged plasterboard.

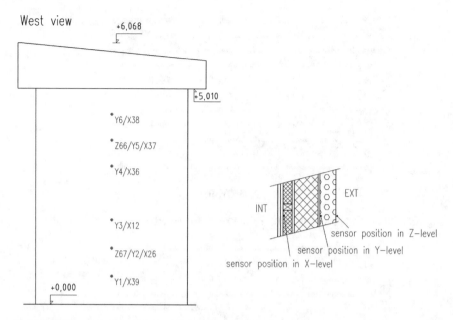

Fig. 3.68 West elevation of the tested building with thermocouple layout and detail of the wall assembly with three different types of thermocouple locations within the wall—plane X, Y and Z

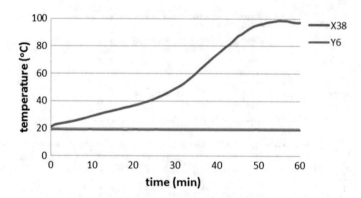

Fig. 3.69 Temperature curves for X38 and Y6 thermocouples

For the other thermocouples in this room, the measured temperature was not higher than 100 °C, although the increase in ceiling temperature (thermocouple S49) was faster than the temperature increase in the beams. It is necessary to keep in mind that a fire is regulated not only by fuel but also by airflow—ventilation. Figure 3.56 shows the course of temperatures near the beams of the first floor with fire protection treatment (see Fig. 3.49 sect. A-A), while the thermocouple N45 was closer to the source of the fire.

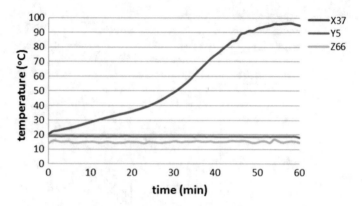

Fig. 3.70   Temperature curves for X37, Y5 and Z66 thermocouples

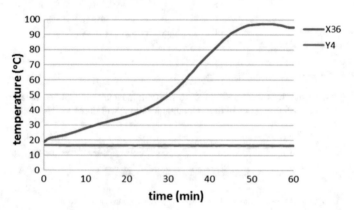

Fig. 3.71   Temperature curves for X36 and Y4 thermocouples

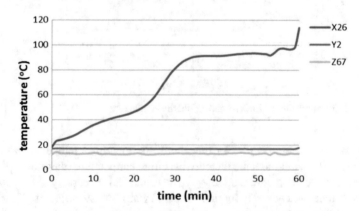

Fig. 3.72   Temperature curves for X26, Y2 and Z67 thermocouples

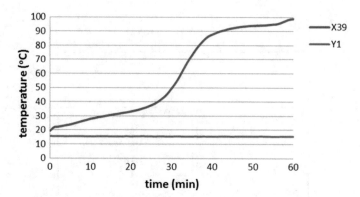

Fig. 3.73   Temperature curves for X39 and Y1 thermocouples

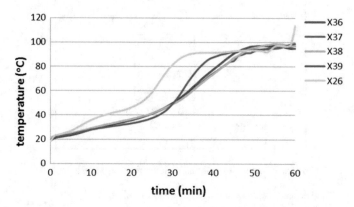

Fig. 3.74   Temperature curves for X36, X37, X38, X39 and X26 thermocouples

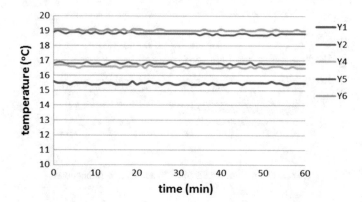

Fig. 3.75   Temperature curves for Y1, Y2, Y4, Y5 and Y6 thermocouples

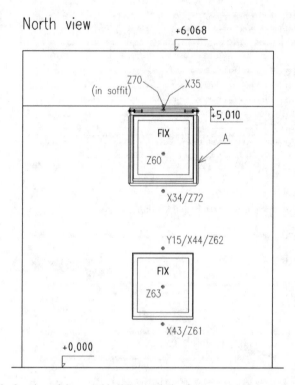

Fig. 3.76 North elevation of the tested building with the thermocouple layout

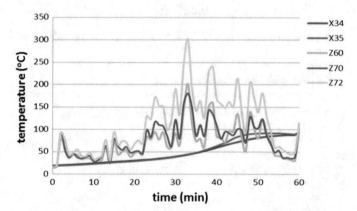

Fig. 3.77 Temperature curves for X34, X35, Z60, Z70 and Z72 thermocouples

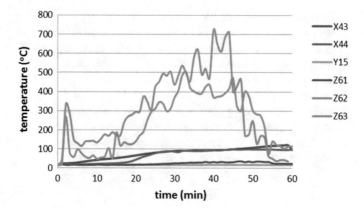

Fig. 3.78 Temperature curves for X43, X44, Y15, Z61, Z62 and Z63 thermocouples

Fig. 3.79 South elevation of
the tested building with the
thermocouple layout

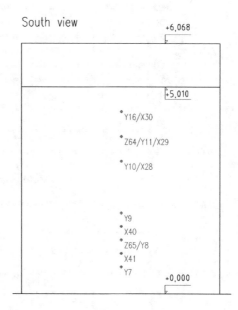

3.5.2 Fire Assessment—Upper Floor

Figure 3.57 shows the positions of thermocouples N19, N20, N27, and N42, which
measured the temperature of the beams on the upper floor. The course of these curves
is shown in Fig. 3.58.

Figure 3.57 shows the location of thermocouples on the ceiling beams in the upper
floor. Thermocouples N19 and N20 were treated with fire protection fabric Protex
110A2, and thermocouples N27 and N42 were treated with fire protection fabric
Heliotex EW90 (see sect. A-A Fig. 3.49).

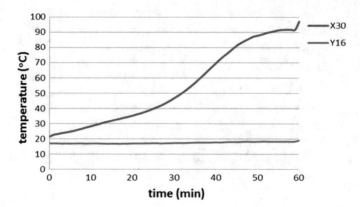

Fig. 3.80   Temperature curves for X30 and Y16 thermocouples

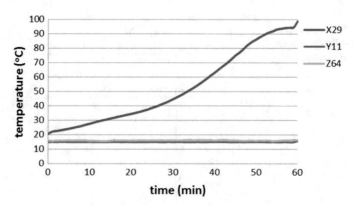

Fig. 3.81   Temperature curves for X29, Y11 and Z64 thermocouples

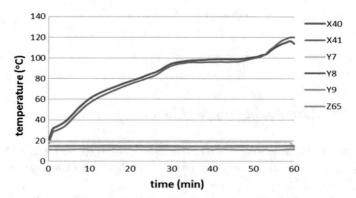

Fig. 3.82   Temperature curves for X40, X41, Y7, Y8, Y9 and Z65 thermocouples

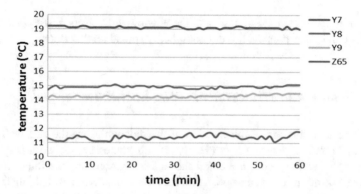

Fig. 3.83 Temperature curves for Y7, Y8, Y9 and Z65 thermocouples

We see that the temperature range is up to 24 °C and the differences between the individual treatments are very small. All materials more or less copy the temperature of the fire flame; the increase begins in the 40th minute. We observe an increase from 18 to 24 °C in thermocouple N20 and from approx. 15.5 to 19 °C in thermocouple N27. If we imagine that a temperature of 1000 °C was reached just a few meters lower, these data testify to the quality of the materials used and the quality of the craftsmanship. It should be noted that the N19 thermocouple stopped measuring at the 40th minute, hence the decrease in the curve. The thermocouple may have simply stopped working, or mechanical damage may have occurred between the thermocouple and the control panel.

The temperature of the ceiling on the upper floor was measured by thermocouples S21 to S25, shown in Fig. 3.59 (position of thermocouples) and 3.60 (temperature course). Thermocouples S21, S24, and S25 were placed under the first plasterboard layer and S23 and S22 under the other claddings.

Thermocouple S25 measured the highest temperature, as it was located closest to the opening that represented the staircase.

Thermocouple S24 was also in a straight line of flow and S21 was moved slightly backward, closer to the external wall. This explains the small differences in the measurement of these thermocouples.

The course of the curves in Fig. 3.60 shows that each layer of plasterboard is important. While the thermocouples under the first layer recorded a temperature in the range of 74–82 °C, the thermocouples under the next layer, directly on the OSB board, did not exceed 25 °C. According to Eurocode 5, it is the thickness of the cladding, its layering, and installation that make it possible to control the fire resistance.

As in the previous cases, the color-coding of the thermocouples corresponds to the color-coding of the temperature curves.

If we were to comprehensively evaluate the fire on the upper floor, the maximum temperature measured was 82 °C in the 60 min of the experiment at one place with nine thermocouples. We must also point out that the window on this floor was protected by the Fireshield system, and the window did not open, the glass did not break (or crack) and thus the fire did not quickly spread on that floor.

3.5.3   Temperature Assessment—The East External Wall

Figure 3.61 is a diagram of the location of thermocouples on the east wall of the building. From this diagram it is clear in which planes and at what heights the thermocouples were placed. The detail shows the location and marking of the thermocouples: position X is on a wooden grid behind the plasterboard cladding from the interior side, position Y is on the OSB board in front of the insulation, and position Z is from the exterior side.

The time dependence of the temperature course measured by individual thermocouples is shown in Figs. 3.62 and 3.63. The thermocouple temperature values from above X33 and Y18 are shown in Fig. 3.62. Thermocouple temperature X33, i. e. behind the plasterboard in the upper part of the building, did not exceed a temperature of 100 °C which is just above 90 °C. Thermocouple Y18, positioned in front of the thermal insulation, had a temperature of about 15 °C the whole time, which actually copies the ambient temperature (the thermometer checking the exterior temperature showed a temperature of 13.5° C). Figure 3.63 presents the results from the second position from above, i.e., thermocouples X32, Y17, and Z68. The temperature course is similar to the previous figure for thermocouples X32 and Y17.

The increase in temperature on the Z68 curve is caused by smoke penetrating through the open door and locally heating the Z68 thermocouple. The peaks in the curve were caused by the airflow, which was regulated by the fire and the airflow in general. Figure 3.64 shows the temperature for thermocouples X31 and Y15 in the third position from above, just above the ceiling of the ground floor. The course of temperatures is almost the same as in Fig. 3.62. The interior temperature is below 100 °C.

Figure 3.65 represents the measurement on the ground floor (where the fuel was ignited), showing the curves of the thermocouples X47, Y13, and Z69 (lower position—next to the door Fig. 3.61). The temperature in position X behind the plasterboard cladding is about 20–25 °C higher compared to the upper floor, approaching a value of 120 °C. This is logical because this wall was directly heated from the interior by the fuel-burning fire. However, the temperature in the Y position did not change and the outcome for the Z position is similar to that in Fig. 3.63, although no major changes are observable on the outer thermocouple due to its position, as is the case with the exterior thermocouple above the door.

Overall, based on the results from thermocouples X of the east wall (X31, X32, X33 and X47 Fig. 3.66) and thermocouples Y of the east wall (Y13, Y15, Y17 and Y18, Fig. 3.67) we can see that the highest temperature was measured by thermocouple X47, which was located on the ground floor. The other thermocouples located on the upper floor recorded an almost identical course of temperatures, as the upper floor space was homogeneously warmed by fuel burning on the ground floor.

As for the thermocouples in the Y position, the highest temperature was measured by the thermocouple Y15, which was behind the insulation above the door; we assume that it was heated by smoke from the outside, but still, its maximum temperature did not exceed 19 °C. The difference between the other thermocouples was only 1 °C.

Figure 3.74 is a comparison of the temperatures of all thermocouples in the X position of the west wall (X36, X37, X38, X39, and X26) where the highest temperature was reached by the thermocouple X26, which was closest to the fuel source in the middle of the ground floor, with an increase in temperature in about 25th minute to 60 °C. Overall it maintains higher values until the final stages of the fire when the values stabilize. The thermocouples in the Y position show different temperatures—the highest for Y2—even if the temperature difference between them is only 4 °C (Fig. 3.75).

3.5.4 Temperature Assessment—The North External Wall

Figure 3.76 represents a measurement scheme, the position of thermocouples on the north wall. We evaluate the measurements of the upper window and the lower window separately. The upper window was measured by thermocouples X34, X35, Z60, Z70, and Z72 (for temperature courses see Fig. 3.77). For thermocouples in position X, which were under the cladding, the temperature did not exceed 100 °C. Thermocouples in the Z position recorded the temperature of the hot gases and the heat radiation from the lower window onto the exterior cladding. The thermocouple in the Z72 position, which was closest to the lower edge of the window, recorded a maximum temperature of 300 °C, in the 30th minute.

The lower window measurements were recorded by thermocouples X43, X44, Y15, Z61, Z62, and Z63. Figure 3.78 shows that the thermocouples in the Z position in the window and above the window were heated by the radiant heat of the fire. In the case of the Z63 thermocouple, which was placed in the window, the temperature exceeded 700 °C. The temperature of 500 °C was measured in the 33rd minute by the Z62 thermocouple placed just above the window. The Z61 thermocouple in the Z position under the window measured a slightly elevated outdoor temperature: the maximum temperature measured was 37.5 °C in the 47th minute. The thermocouples in the X position (X43 and X44) only slightly exceeded 100 °C and the thermocouple in position Y did not exceed 20 °C.

3.5.5 Temperature Assessment—The South External Wall

The exterior thermocouple layout on the south external wall of the building is shown in Fig. 3.79.

Both the structure of the marking and the description of the positions and evaluation of the temperature course at the position of the thermocouples from top to bottom are kept uniform.

With the first-floor thermocouples X30 and Y16 (see Fig. 3.80), the temperature measured by thermocouple X30 did not exceed 100 °C, and thermocouple Y16 measured a temperature below 20 °C. The second measurement from above is represented by thermocouples X29, Y11, and Z46 (Fig. 3.81), and again, the temperature did not exceed 100 °C. Thermocouples Y11 and Z64 measure almost identical temperatures around 15 °C.

Upper-floor thermocouples X40, X41, Y7, Y8, Y9, and Z65 are shown in Fig. 3.82. As with the previous walls, the X40, and X41 thermocouples reached a temperature of 120 °C on the ground floor, as they were located in the area that is in direct contact with the fuel. Thermocouples Y7, Y8, Y9, and Z65 measured a temperature of about 20 °C. For clarification, we have changed the range of the y-axis, which represents the temperature in Fig. 3.83 by omitting the thermocouples in position X from 140 to 20 °C; see Fig. 3.83, where you can better see the temperature differences between thermocouples Y7, Y8, Y9, and Z65. As can be seen from the figure, they all measure temperatures below 20 °C.

Reference

1. Osvald A (2011) Drevostavba ≠ požiar. (A timber structure ≠ a fire). Technická univerzita vo Zvolene, Zvolen. pp 336. ISBN 978–80–228–2220–6

Chapter 4
Conclusion–Summary of the Experiment

Abstract The chapter contains a summary of the course of the large-scale fire test of a two-story wooden building built expressly for the purpose of this test.

Keywords Two-story wooden building · Model fire · Large-scale fire tests

The experiment is described in detail in Chap. 3. To evaluate the fire test of the reference building, it must be stated that the test was financially challenging. Only with the help of the Slovak Association of Wood Processors were there several construction companies willing to participate in the experiment.

The overall preparation of the model test was both time-consuming and expensive. In the first phase, a project was prepared in which recalculations were performed according to Eurocode 5, and only materials that had a valid certificate were used. We assumed that it would be a homogeneous building, where one thermal insulation material and one fireproof cladding material would be applied, and thus the building will be similar to a real building.

Since several companies took part in the experiment, it was logical that each of them wanted to use their own materials. This caused several problems at first, but we eventually accepted this request and found a compromise. This way, a building where each wall was different, and each half of the ceiling had a different thermal insulation material (see one of the photos), was created. We were afraid that such an atypical building might not withstand fire. However, the result was surprising.

Quality testing of materials and quality craftsmanship proved that even with such a "mixture" of materials applied in the building, a fire-resistant building model could be prepared. Evaluation of the experiment according to individual materials turned out to be unnecessary, as we would only advertise individual materials, and we asked ourselves the question: "Is such a detailed comparison of materials necessary if the temperature on the opposite side between the individual materials recorded a difference of only 2–3 °C?" For this reason, we abandoned such an evaluation. We can reliably state that if the material has a valid certificate and valid tests performed according to the new regulations (reaction to fire and fire resistance), it can be applied in the construction and is reliable in terms of fire protection.

The experiment proved to be justified. It proved that wooden buildings can with-stand fire and they can be safe in the event of an emergency such as a fire. Certification is crucial, but it alone is not enough. Both a quality project taking into account all the fire protection measures (escape routes, distances, etc.) and Eurocode 5 calculations are necessary. These allow us to determine, in a professional way, the fire resistance of a building.

Printed in the United States
by Baker & Taylor Publisher Services